Acoustic Emission

Nondestructive Testing Monographs and Tracts
A series edited by Warren J. McGonnagle

Volume 1
Microwave Nondestructive Testing Methods
by Alfred J. Bahr

Volume 2
Acoustic Emission
Edited by James R. Matthews

Additional volumes in preparation.

ISSN: 0730-7152

This book is part of a series. The Publisher will accept continuation orders, which may be cancelled at any time and which provide for the automatic billing and shipping of each title in the series upon publication. Please write for details.

Acoustic Emission

Edited by

James R. Matthews
Defense Research Establishment
Atlantic Dockyard Laboratory
Halifax, Nova Scotia

Gordon and Breach Science Publishers
New York London Paris

Gordon and Breach, Science Publishers, Inc.
One Park Avenue
New York, NY 10016

Gordon and Breach Science Publishers Ltd.
42 William IV Street
London WC2N 4DE

Gordon & Breach
58, rue Lhomond
75005 Paris

Library of Congress Cataloging in Publication Data

Main entry under title:

Acoustic emission.
(Nondestructive testing monographs and tracts
ISSN 0730-7152 ; v. 2)
Includes index.
Contents: Acoustic emission evaluation / James R. Matthews and D. Robert Hay – Acoustic emission from dislocation motion / C. R. Heiple and S. H. Carpenter – Acoustic emission signal characterization / Ronald A. Kline – [etc.]
1. Acoustic emission–Addresses, essays, lectures. 2. Acoustic emission testing–Address, essays, lectures. I. Matthews, James R. II. Series.
TA418.84.A257 1983 620.1′127 82–20928
ISBN 0-677-16490-4.

ISBN 0-677-16490-4.
ISSN 0730-7152.
Printed in the United States of America.

Contents

Introduction to the Series

This series concentrates on the methods of and the techniques in nondestructive testing (NDT), and will publish information and data obtained from: applications, experimentation and theory; improved methodology; instrumentation modification and developments; application of the methodology to new and different problems or problem areas, and analysis, discussion and evaluation of standard methodology. Each volume may be devoted to one method or subject or may contain papers on a number of methods.

Associate editors will assist in the selection of the subject matter, as well as in evaluation of manuscripts submitted for publication in the series.

We believe these monographs and tracts will help advance the science, engineering and technology of NDT and stimulate its application to industrial problems, thus helping to assure the quality, reliability and safety of industrial materials, components and systems. Some of the volumes will also be useful in seminars and courses in NDT materials and related areas.

Comments concerning this series are welcome.

Warren J. McGonnagle

Acoustic Emission Evaluation†

James R. Matthews
Defense Research Establishment
Atlantic Dockyard Laboratory
FMO Halifax, Nova Scotia B3K 2XO

and

D. Robert Hay
Tektrend International Inc., 750 Belair St.
Montreal, P.Q., H4C 2K3

1. INTRODUCTION

Acoustic Emission[1] (AE) is a term describing a class of phenomena whereby transient elastic waves are generated by the rapid release of energy from localized sources within a material. Several other terms used to describe these phenomena have included stress-wave emission, microseismic activity and microseismic emission. AE has developed rapidly over the last two decades as a nondestructive evaluation technique and as a tool for materials research. It is a high sensitivity technique for detecting active microscopic events in a material. Since these events give rise to elastic waves which propagate out into the material, it is not necessary to focus in on the exact location of the source event to detect it. Thus AE is a passive monitoring technique which, rather than requiring 100% volumetric scanning (such as in radiography, ultrasonics, or eddy currents), can inspect using remote sensors. It, therefore, can be more efficient with respect to preparation and inspection time, considerably reducing costs. In many applications it is also more sensitive than other nondestructive testing (NDT) techniques. It is often this increased sensitivity which is exploited by using it to determine areas of a structure requiring more detailed characterization by other NDT methods. However, work is now being carried out actively to develop the ability to characterize the source from the waveform put out by the AE sensor.

The first documented observation of acoustic emission in 1953 by J. Kaiser[2] was the modern day starting point for its active use. In fact, Kaiser characterized a basic irreversibility phenomenon which bears his name. In the Kaiser effect (Figure 1) when a material is stressed to a given level and the stress

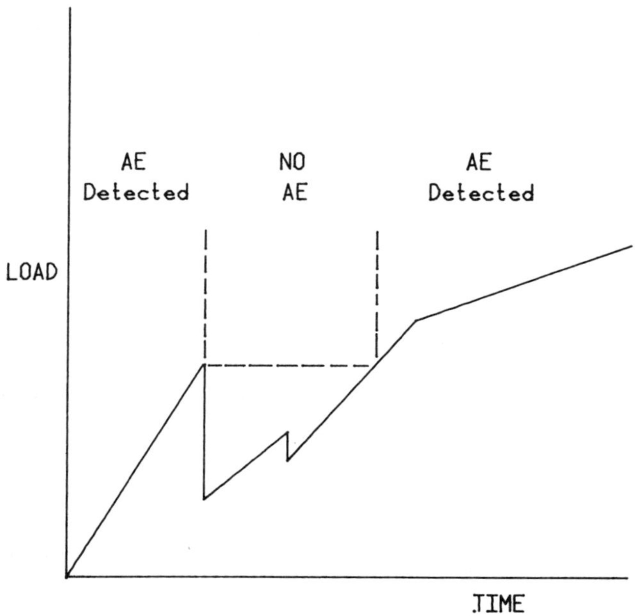

Figure 1. Kaiser effect.

removed, upon reloading there is an absence of detectable emission at a fixed sensitivity level until previously applied stress levels have been exceeded. Since Kaiser's time, advances in both materials science and electronics technology have contributed to bringing acoustic emission to the forefront of new NDT methods.

In the past, McGonnagle[3] has suggested that new physical phenomena be evaluated for their potential as nondestructive testing tools and that resolving power and sensitivity of all methods be increased and new instrumentation be developed. Acoustic emission testing is a response to such technological demands. This article focuses on the efforts to utilize the phenomenon for nondestructive testing to improve its resolving power and sensitivity and to enhance the ability to interpret data through developments in instrumentation and analysis methods. To start, the nature of flaws in various types of materials will be outlined. This will be followed by general descriptions of current instrumentation, applications and new signal processing methods.

2. NATURE OF FLAWS AND NATURE OF AE IN VARIOUS MATERIALS

Acoustic emission testing requires that flaws be excited under an applied load, either a proof load or an operating load. Thus, the nature of the acoustic emission is intimately related to the types and sizes of flaws and their growth mechanisms.

Critical defect sizes in brittle materials are among the smallest of all materials of the order of 1μ to 50μ depending upon the application and the material. Polycrystalline KCl, for example, has a fracture toughness of 200 MPa $\sqrt{m}$ and a yield strength of 600 kPa giving a critical defect size the order of the grain size. In silicon nitride structural materials, the critical defects are silicon inclusions around 25μ in size. Finding these flaws by ultrasonics[4] is a process of great effort. However, due to their size, the prospect of localized proof-testing while listening with acoustic emission may offer the possibility of a practical inspection method.

In metals, significant defects can range from a few mm up to a meter. Propagation of these defects[5,6] can be preceded by crack blunting and local yielding, crack advance in ductile tearing, or crack advance associated with rapid release of energy governed by presence of hard particles in the metal structure. The latter may involve fracture of the larger particles or pullout of the smaller ones. The reverse is also true for porous materials where massive strain occurs around the defects as they grow and coalesce giving rise to a significant rate of acoustic emission.

In composite materials, there may be a multiplicity of defects and failure modes. The defects of note result from fibre strength variability, matrix quality, and the adherence of matrix to fibres. Consequently, efforts go into utilizing AE to differentiate between the modes of failure implied here, namely, fibre fracture, fibre pullout and matrix cracking.[7,8]

Geological defects which result in such phenomena as earthquakes, volcanoes and land slides come next in the order of Size. An example of the use of AE in detecting impending geological failure is the use of steel rods driven 10 meters into soil in the vicinity of potential slides with standard AE sensors attached to the exposed end. Significantly advanced warning (typically 3–4 weeks)[9] has been provided in many cases.

The microphenomena which give rise to acoustic emission range from those with high detection success potential,[10] such as detection of high energy rate mechanisms during plastic deformation, to the much more difficult such as monitoring and estimating elastic stresses.[11]

Fortunately, in many cases the high energy events are those which are of interest in NDT making it possible to exploit the method for inspection. In general, the different phenomena give rise to signals that are either burst or continuous. The mechanics of acoustic emission, whether continuous or burst, rely upon the release of energy at a continuous source such as dislocation motion or burst source such as crack advance. The released energy propagates as a stress wave to a free surface which then oscillates producing either a pressure on a piezoelectric type transducer or interfering with an optical deflection measuring device. In each case, the pressure or deflection reading is transformed to a voltage time output. It is this transformation of the acoustic emission event which provides for electronic monitoring, signal enhancement analysis, and ultimately, determination of component integrity.

In summary, in order to properly interpret AE signals, it is necessary to fully understand the nature of the flaws in the material under evaluation. Likewise, the relation between their propagation mechanisms and the AE produced is required for interpretation. These relationships are not required when AE is used as a complement to a proof test as the source indications can always be verified using other NDT methods. However, for on-line surveillance of structural components, these relationships must be known before testing to provide a real-time decision making capability compatible with the inherent real-time detection capability of the inspection method.

3. INSTRUMENTATION

Sensors are the backbone of AE monitoring. They come in a variety of packages with trade offs between sensitivity and flat frequency response. It has always been difficult to provide exact calibration of surface deflection under the sensor. The best balance here has resulted in the widespread use of piezoelectric transducers. For laboratory work these transducers are often used side-by-side with laser deflection measurement systems. Although thought has been given to creating miniaturized laser systems for field work, greater expectations are being placed in the successful application of fibre optic transducers.

Calibration of sensors has taken a variety of courses, the

most widespread being the excitation of a transducer with a standard broadband transducer and noting the resulting frequency response. A great deal of effort has gone into using mechanical sources such as fracture of lead pencils and glass rods, calculating the transfer function from the source to the transducer locations, and comparing these to the transducer output. This has also been done in conjunction with laser measurement. To take advantage of enhancement possibilities in the time or frequency domain, calibration and matching of transducers in both domains is a must.

Preamplification is necessary to limit noise problems in AE field work. AE manufacturers have gone so far as to provide certain transducers with preamplification built in. The typical preamplifier is 60 dB (1000X) giving an output around 10 volts peak-to-peak. These preamplifiers are typically excited by ± 12v dc drawing about 20 milliamps and in noisy field and delicate laboratory environments it is wise to provide this from batteries. Most general purpose equipment is provided with a typical bandwidth of 100 kHz to 2 MHz but 45 kHz to 4 MHz is common. A large variety of standard filter combinations is available for use in these bandwidths for specific applications.

Many self-contained products are available for providing basic AE output. These include devices for threshold counting (ring down), rms signal level monitoring, count rate monitoring, basic two channel flaw locating, and multichannel (16–64) monitor systems designed for pressure vessel proof testing and refinery or nuclear power plant monitoring.

In the laboratory and field areas, a convenient recording device for individual signals is the digital oscilloscope which today can be obtained with sampling rates up to 20 MHz. These devices accept two inputs simultaneously and are available with numeric cursor control. They can be connected directly to mini computers with standard IEEE-488 interfaces. From there on a wide range of digital signal processing techniques can be used to isolate signals from noise and to define basic signatures.

Continuous recording of AE signals is extremely difficult and expensive whether digital or analog. Analog recording is possible for up to 10 minutes duration with up to 35 dB signal-to-noise ratios for multiple channels. Due to the high frequencies involved, it is not reasonable to expect to record long tests either in the laboratory or in the field. Direct data observations and snap shot analysis are accordingly the most widely used research methods. For large programs, however, such as structural aircraft monitoring, scheduled recording is necessary considering the high cost of a flight test and the cost of fatiguing an aircraft in a ground test.

4. APPLICATIONS

Proof testing, whether of pressure vessels,[12] nuclear reactor piping,[13] castings or aircraft, is one of the major application areas for AE. Many groups are active in this area including inspection companies, government labs and NDT groups in larger industrial firms such as petrochemical and industrial gas companies. Generally flaws are detected and located by placing a network of transducers on the suspect areas followed by application of a test load. This may be a hydrotest of a pressure vessel or a piping system, or flexure test in the case of an aircraft wing bend fixture test. Each channel is monitored in as quiet a test state as possible and sources of emission are located by triangulation. Once found, these areas are re-examined by conventional NDT methods. Data recording and presentation are facilitated by dedicated minicomputer systems.

Acoustic emission monitoring of welding has received considerable attention. Welding is monitored in progress and during the 72 hour period after welds are completed.[14,15] During this period, hydrogen is escaping from the weldments and hydrogen cracking is possible.[16] In assembly line situations, weld monitoring devices have successfully been developed which vastly reduce inspection costs, particularly in the case of spot welds.[17] Continuous weld monitoring devices for large structures have also been developed.[18] AE has also been used as a quality monitor in rolling, punching, and machining operations.[10]

The application of AE to geological structures[19,20] has received widespread attention. The structures looked at have included mines, highways (slide areas), water cuts, dams, petro reservoirs, and underground gas storage. The many tunnels constructed prior to the 1950's and still in use are a concern to highway authorities and use of AE to guarantee integrity has become an important activity.[21]

A final important application is in assurance of structural integrity by requalifying older structures. Many successful applications have been made to bridges, wooden beams, pipelines, older buildings and, of course, aircraft[22,23] and nuclear power plants.[13] It is the latter areas where the impetus for the recent efforts in signal processing outlined in the next section was developed.

5. SIGNAL PROCESSING

Acoustic emission signals can take many forms dependent upon the material and the failure mechanism in the material. AE signals from defects in composites and geological materials generally contain information at low frequencies .5–100 kHz. Attenuation is high due to the complexity of the material. Signals of significance in metal and brittle materials contain information between 100 kHz and 2MHz. In this range a good compromise is found for most testing applications because ambient noise is low and the cost of electronic components is reasonable.

If there are many types of AE signals there are also many sources of noise which can interfere with detection of the critical event. Noises include the hydraulic noise of flowing fluids, fretting, vibration or impacting of loose parts, and general plant or field background noise. A number of techniques are employed to isolate valid signals from noise in the time domain. Times of arrival can be used to permit geometric elimination of obvious noise through gating and the acceptance of only those signals which arrive from a particular region of the structure. Also, characteristic signals can be compared to those produced in laboratory analysis.

In the frequency domain extensive efforts have also been directed at differentiating between signals and noise. One approach involves subtracting the frequency response of a known source introduced in the same location as the expected failure on the actual structure or component. However, due to the dominant fraction of the frequency content contributed by the micro and macro structural characteristics such as grain size and part configuration, the tranducer and the couplant, the contributions from the source event are often overwhelmed and lose prominence in the spectra. This method, in addition, does not allow for absolute calibration.

Amplitude distribution and duration distribution are extensively utilized in AE processing, Figure 2. These are also two

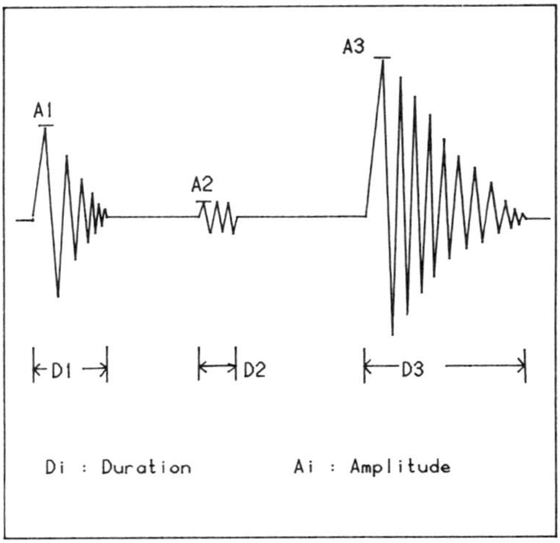

Figure 2. Amplitude and duration distribution. The above figure shows measurement of amplitude and duration of unknown events. Frequency of occurrence histograms are plotted from these valves.

of the basic features used in pattern recognition in the time domain. A simplistic approach to pattern recognition in the time domain is shown in Figure 3 where several additional signal features are defined.

Pattern recognition is of most value when the noises and signals do not occur simultaneously (Figure 4). In this case, since the signals are separated in time, their pure nature can be examined and interpretation can lead to isolation of fracture problems. This can be done in real time by a specialized monitor or periodically in a laboratory by examination of recorded data tapes. The percentage of quiet time monitored by a transducer is a good indication of the success probability of a pattern recognition approach. With minimal quiet time the probability of occurrence of noises and signals together is large, whereas with a lot of quiet time the probability of the two sources acting together is small.

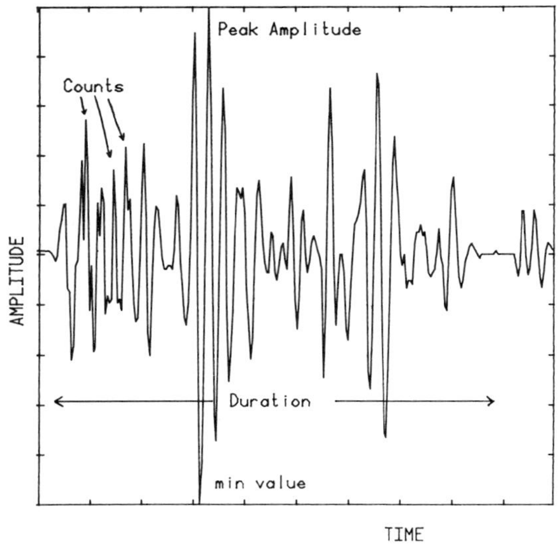

Figure 3a. Pattern recognition in the time domain. Peak amplitude, minimum value, area, rise time, locations of peak and minimum amplitudes, duration, location of maximum slope, mean value, RMS, Standard deviation, variance, counts above specific threshold.

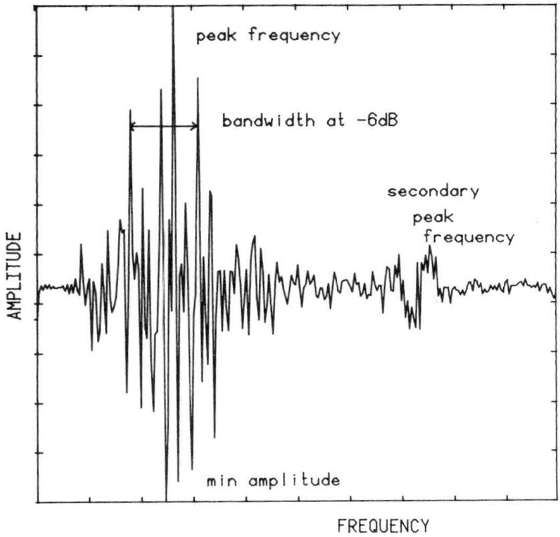

Figure 3b. Pattern recognition in the frequency domain. Peak frequency, maximum amplitude, minimum amplitude, area, maximum slope, frequency at maximum slope, frequency at minimum amplitude, 6db down band width, mean value, secondary peak frequency, RMS, standard deviation, variance.

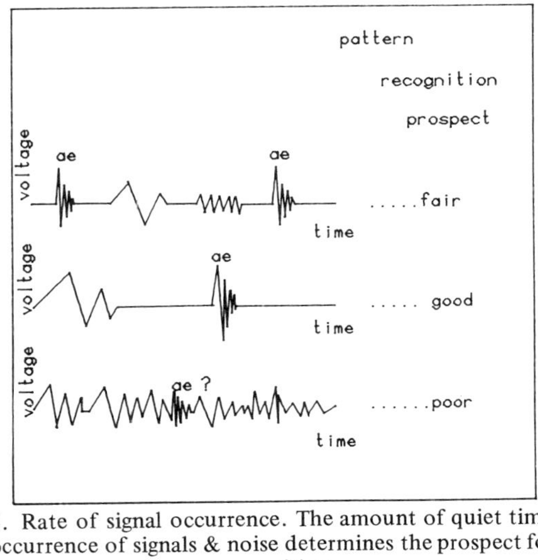

Figure 4. Rate of signal occurrence. The amount of quiet time and the rate of occurrence of signals & noise determines the prospect for pattern recognition.

Correlation methods, advanced transformations, adaptive learning and other signal enhancement techniques will be required, however, for the successful isolation of cracking phenomena in continuously noisy structures. This will be the focal point of some of the following papers.

In the area of interpretation, much effort is concentrated on the use of stand-alone computers and other dedicated electronics. For 100% reliability, signal-to-noise ratios of 35 dB are often felt to be required by electronics designers. This may be difficult to achieve in many important AE applications and it may be that some computer-operator-expert interaction is required, especially in the learning process.

Last, but not least, much can be achieved in the signal/noise ratio battle by effective use of gating techniques, and planned loading sequences for intermittent monitoring of service equipment. The added information gained or eliminated by the selective use of a well-defined recent loading history can make the difference in detecting growing defects. In a flight test, for example, a one hour flight in calm weather without manoeuvers can serve to sharpen fatigue cracks prior to a scheduled AE monitoring period during well defined manoeuvers. In general, a monitor that merely listens may be ineffective but its usefulness increases dramatically if one can periodically control or apply a given loading sequence.

Deconvolution of specific AE signals with libraried data from known sources might also prove useful in identifying AE sources by the pattern recognition approach or be useful in enhancing background signals when continuous noise is a problem. This technique is difficult because its use must be tailored to specific applications and the method quickly becomes unstable.[24,25,26]

6. SUMMARY

Since its latter day discovery in 1953, our understanding of AE, its relationship to the mechanisms which produce it, and the way in which to apply it for nondestructive evaluation has progressed dramatically. It has found acceptance in several well-defined applications and the range of future potential applications is broad. This will be expanded as further research and development work proceeds in key areas such as transducer technology, and signal analysis.

REFERENCES

1. ASTM E07.04 Subcommittee on AE, "Standard Definition of Terms Related to AE", 1980.
2. J. Kaiser, Untersuchungen Uber das Auftreten Gerauschen Beim Zugversuch Archiv fir das Eisenhüttenwesen, Vol. 24, 1953.
3. W. J. McGonnagle, "Non Destructive Testing" in Fracture, edited by H. Liebowitz, Vol. III, 1971.
4. A. G. Evans, G. S. Kino, P. T. Khuri-Yakub and B. R. Tittmann, "Failure Prediction in Structural Ceramics" Materials Evaluation, pp. 85–96, 1977.
5. F. A. McClintock, "Plasticity Aspects of Fracture" edited by H. Lieboweitz, Vol. III, 1971.
6. A. S. Tetelman, "Acoustic Emission Testing and Microfracture Processes" Materials Res. Stand, VII, 13–16, 1971.
7. C. D. Bailey, S. M. Freeman, J. M. Hamilton Jr., "Acoustic Emission Monitor Damage Progression in Graphite Epoxy Composite Structure" Material Evaluation, Vol 58, #8, August 1980.
8. R. G. Liptae, "Acoustic Emission from Composite Materials", STP 497, ASTM, Philadelphia, 1972.
9. Al Green, Personal Communications.
10. Robert C. Liptae, Clement A. Tatro, "Acoustic Emission Inspection" in Metals Handbook, Vol. II, 8th Edition, ASM, Metals Park, Ohio 44073, 1976.
11. J. H. Williams, S. S. Lee, "Monitoring of Elastic Stresses by AE" Dept. of Mechanical Engineering, Massachusetts Institute of Technology.
12. K. Stahlkopt, A. Green, "Non Destructive Pressure Vessel Testing by Acoustic Emission", EPRI Journal, Vol. 1, #1, 1976.
13. S. K. Edler, Ed. Reactor Safety Research Programs Quarterly Reports prepared by Pacific Northwest Lab for U.S. Nuclear Regulatory Commission.
14. W. D. Jolly, "The Application of Acoustic Emission to In-Process Inspection of Welds", Materials Evaluation, Vol. 28, #6, June 1970.

15. M. N. Bassim, D. R. Hay, M. Talerman, "Postweld Evaluation by Acoustic Emission", Materials Evaluation, Vol. 37, #6, 1979.
16. K. Masubuchi, Analysis of Welded Structures, Pergamon Press, Toronto, 1980.
17. M. C. Jon, C. A. Keskimaki, S. J. Vahaviolas, "Testing Resistance Spot Welds Using Stress Wave Emission (SWE) Techniques", Materials Evaluation, Vol. 36, #4, 1978.
18. A. E. Wehrmeister, "AE Monitoring of Multi-Pass Submerged Arc Welding", Materials Evaluation, Vol. 35, #6, 1977.
19. A. E. Lord Jr., R. M. Koerner, W. M. McCabe, "Acoustic Emission Behaviour of Sand as used in Foundation Bearing Capacity", Materials Evaluation, Vol. 34, #5, 1976.
20. H. R. Hardy, F. W. Leighton, editors "Proceedings of 1st Conference on Acoustic Emission/Microseismic Activity and Geologic Structures and Materials", Penn State University, June 9–11, 1975, Transtech publication, 1977.
21. Acoustic Emission Working Group, Personal Communications.
22. W. M. Pless, C. D. Bailey, J. M. Hamilton, "AE Detection of Fatigue Crack Growth in a Production-Size Aircraft Wing Test Article Under Simulated Flight Loads", Materials Evaluation, Vol. 36, #5, 1978.
23. C. D. Bailey, W. M. Pless, "Acoustic Emission Structure-Borne Background Noise Measurements on Aircraft During Flight", Materials Evaluation, Vol. 34, #9, 1976.
24. K. Fukunaga, Introduction to Statistical Pattern Recognition, Academic Press, NY, 1972.
25. R. O. Duda, P. E. Hart, Pattern Classification and Scene Analysis, John Wiley & Sons, Toronto, 1973.
26. V. L. Newhouse, E. S. Furgason, "Random Signal Flaw Detection", Purdue University, AFML-TR-79-4077, June, 1979.

Acoustic Emission from Dislocation Motion

C. R. Heiple
Rockwell International
Rocky Flats Plant
Golden, Colorado

and

S. H. Carpenter
Department of Physics
Denver University
Denver, Colorado

1. INTRODUCTION

Acoustic emission is a transient elastic wave, large enough to be detected, which is generated by the rapid release of energy within a material. Numerous mechanisms have been proposed and confirmed as sources of acoustic emission, however, this review is concerned only with acoustic emission generated by dislocation motion. Experimentally, the majority of acoustic emission from dislocation motion during deformation generally occurs near the onset of plastic flow. Twinning and martensitic phase transformations are excluded from this review even though dislocation motion is involved in the nucleation and growth of twins and the growth of martensite. The authors have reviewed acoustic emission from dislocation sources previously,[1] and several recent, more general, reviews of acoustic emission are also available.[2-5] Within the last few years there has been substantial progress in understanding the nature of acoustic emission from dislocation motion and the effects of various material and experimental variables on the observed acoustic emission. We believe an updated review is therefore appropriate.

2. EXPERIMENTAL TECHNIQUES

The success of any experimental investigation rests on the ability to first extract from the experiment meaningful and useful data, and second, to analyze and format the data to provide information pertinent to the goals of the investigation. Acoustic emission phenomena are complicated and difficult to decipher unless precautions and considerable experimental skill are employed throughout the investigation. Questions which must be considered in some detail include:

1) What is the proper experimental approach?
2) What is the best equipment, and what type of signal conditioning should be employed?
3) What type of analysis will provide the most insight into the particular questions being investigated?

2.1 Continous and Burst Emission

Acoustic emission signals have been conventionally separated into burst-type and continuous-type emission. If the signal consists of pulses detectable from background noise and well enough separated in time so that there is not a lot of overlap, then the emission is called burst emission. If resolution of individual pulses is not possible, then the emission is called continuous emission. The authors have long maintained, in agreement with many others, that the only difference between continuous and burst emission is that the bursts in what appears to be continuous emission are too small to be individually resolved. In part, this belief stems from the lack of a plausible mechanism for anything other than burst production during deformation. There is evidence which demonstrates that the signal responsible for continuous-appearing emission is a collection of pulses rather than a sinusoidal oscillation with superimposed noise. Hatano has determined the autocorrelation function of the continuous-appearing emission from 1100 aluminum in tension.[6] The autocorrelation function is the

Fourier transform of the power spectral density of the observed signal. Standard statistical analysis shows that the autocorrelation function for the signal from a process producing rectangular pulses at random and independent times with an average density λ is of the form $\exp(-2\lambda|t|)$, where t is time. On the other hand, if the basic signal is a sine wave with superimposed noise, the autocorrelation function is oscillatory.[7] The autocorrelation function determined by Hatano for the acoustic emission signal from 1100 aluminum is monotonically decreasing and is approximated by an exponential decay with time.

2.2 Sensors

The sensor used to detect acoustic events is of critical importance in any acoustic emission experiment. In the majority of investigations, a piezoelectric transducer is attached directly to the sample being tested. The output of the transducer (electrical voltage) is then amplified, possibly filtered, and processed in any number of signal conditioning schemes. Since the first element in any signal processing scheme or analysis will be the transducer (sensor), considerable care must be exercised in the choice and placement of the transducer. The most common active element has been a piezoelectric crystal of the PZT-5 type. Both resonant- and damped-type transducers have been used. The resonant type provides a high sensitivity at the resonant frequency, while the damped gives a more or less flat (and less-sensitive) response over a wider frequency range. The piezoelectric transducers provide practical, low cost, and reproducible sensors which are readily available and easy to work with. The usual piezoelectric transducer output, however, does not give an accurate description of the displacement of the sample surface. In different parts of its operating bandwidth, a piezoelectric transducer may be sensitive to surface displacement, surface velocity, and perhaps surface acceleration.[3]

Capacitive[8,9] and optical[10] sensors have been developed at several laboratories to measure the absolute surface displacements produced by actual or simulated acoustic emission sources. The advantage of this approach is that if the time history of the surface displacement produced by an acoustic emission event is known, then under some conditions the displacement-time history of the acoustic emission source can be calculated. Such calculations have been performed from surface displacements caused by fracture events in specially shaped steel samples[11-13] and for events attributed to dislocation motion in an Al-Zn-Mg alloy.[14] The disadvantage of these sensors is their substantially lower sensitivity compared to piezoelectric transducers. For example, the minimum surface displacement detectable with the capacitive transducer and subsequent electronics system used at Harwell[9] is 10^{-12} m. There have been several calculations[15-17] of the minimum displacement detectable by piezoelectric transducers leading to predicted sensitivities of 7×10^{-13} m and better. Scruby and coworkers[14,18] have used a pulsed laser to stimulate elastic waves thermoelastically and demonstrated a minimum detectable displacement of about 10^{-14} m using their broadband piezoelectric transducers. Resonant transducers are expected to be somewhat more sensitive.

A piezoelectric transducer with an active element in the form of a truncated cone has been invented at NBS.[19] The truncated end of the cone has a diameter of about 1.5 mm and approximates a point receiver. The shape of the output voltage versus time from this transducer is nearly identical to the output from the NBS standard capacitor transducer from the same transient surface displacement waveform input. The sensitivity appears comparable to that of commercial piezoelectric transducers and it is therefore of potentially great interest for fundamental studies of acoustic emission from dislocation motion. This transducer is not commercially available at the present time, and in its present state of development it would require a specialized testing setup because it is quite fragile and is designed for mounting on a horizontal surface.

The development of new sensors and the calibration and standardization of transducers is an active field of investigation, however, a detailed review of efforts in these areas is beyond the scope of this paper. Considerable progress has been made in the development of new sensors; nevertheless, with the one exception noted above, all reported measurements of acoustic emission from dislocation motion have been made with commercial piezoelectric transducers or similar transducers fabricated by the investigator.

2.3 Signal Processing

Once the acoustic emission events have been detected, it is general procedure to amplify, possibly filter, and then employ some type of signal processing to format and present for analysis the acoustic emission data. At the present time, one can find many presentations of acoustic emission data in the literature. These include: ringdown counts, rms voltage, burst rate, energy rate, rise time, event duration, frequency content, amplitude distribution, and numerous others.

The most common measure of acoustic emission reported in the literature is what is known as ringdown counts. The measurement consists of counting the number of times the amplified acoustic emission signal crosses a preset trigger level voltage. The results can be reported either as a rate, i.e., counts per unit time, or as accumulation (summation) of counts. Harris, Tetelman, and Darwish[20] show that if the voltage output from a transducer excited by an acoustic emission event has the form

$$V = V_o e^{-\beta t} \sin \omega t \tag{1}$$

then the number of counts measured from that event should be given by

$$n = (\omega/2\pi\beta) \ln (V_o/V_t) \tag{2}$$

where V_t is the threshold voltage.

Real bursts are much more complicated than implied by Eq. (1). For example, Esmail and Grabec[21] measured the number of counts from a burst as a function of threshold voltage and found $n \propto V_t^{-5.1}$ for martensite formation in a steel specimen. Egle[22] has proposed that the actual burst may be represented as the product of a decaying exponential and a stationary Gaussian random process with zero mean. He obtained good agreement between his model and bursts arising from crack propagation in a bolt loaded, wedge-opening loaded type of specimen.

For continuous-appearing emission the signal will generally be a narrow band random signal because of bandpass filtering and transducer resonances. The probability density distribution of the peak amplitudes will then have a Rayleigh distribution, and Hamstad[23] has shown that the count rate $\dot{N}$ will be given by

$$\dot{N} = fe^{-V_t^2/\alpha(V_{rms_M})^2} \qquad (3)$$

where f is the apparent frequency of the signal, $Vrms_M$ is the measured root mean square voltage of the signal, and α is a constant equal to 2 for a Rayleigh distribution. The measured rms voltage is composed of both the acoustic emission signal and noise. If the two are statistically independent (and no evidence is available that such is not the case), then

$$V^2_{rms} = (V_{rms_M})^2 - (V_{rms_N})^2 \qquad (4)$$

where V_{rms} is the rms voltage from the acoustic emission signal and $Vrms_N$ is the rms noise voltage.[23] Ono[24] has fit Eq. (3) to narrow band random signals with $\alpha = 2.05$. A good fit to Eq. (3) was also found for experimental acoustic emission data for varying threshold voltages.[25] It is clear that the count rate is very sensitive to threshold voltage and to system gain (through the proportionality of $Vrms_M$ to system gain). The sensitivity of count rate to system gain has also been demonstrated.[26] A good reference paper on the use of ring-down counting has been published by Brindley *et al.*.[27]

In spite of the sensitivity of ringdown counts to system gain and threshold voltage, and the lack of a direct correlation between count rate and event rate, it has been argued that ringdown counts, as a practical measure, is one of the most meaningful measurements of acoustic emission activity because it gives a pseudonatural increased weighting to larger events or bursts. Large events are generally regarded as being most significant, hence, large count values are associated with severe material damage.

In many acoustic emission investigations, the rms voltage generated by the acoustic emission signals is reported as the measure of the acoustic emission activity. Numerous investigators feel that the rms voltage as a measure of the acoustic emission activity has a significant number of advantages over other data formats. The rms voltage measurement is simple, without electronic complications. It does not depend on any type of trigger level, but rather measures all of the emissions generated with equal treatment given to all signals. Furthermore, it can easily be related to the energy of the acoustic emission signals. Hamstad and Patterson[28] have shown, by inputting tone bursts of known energy into an rms voltmeter, that the impulse response of the meter as measured by the maximum deflection of the meter for a given input pulse is related to the energy in the pulse. The relation is

$$E = K_1 K_2 V_{rms}^{K_3} \tag{5}$$

where K_1, K_2, and K_3 are constants depending on the system, the rms meter, and the scale used on the rms meter. Comparisons between count rate and rms voltage are difficult and must be done with care because of the exponential dependence of count rate on $Vrms_M$, as shown in Eq. (3).

Commercial instrumentation which gives an energy measurement depends on the proportionality between the energy contained in the acoustic emission signal and the integral of the square of the rms voltage over the time of interest.

In those situations where the acoustic emission activity is primarily due to individual bursts the acoustic emission might be characterized in terms of the number or rate of bursts, the amplitude of the bursts, the rise time, or the burst duration. Presentation of the acoustic emission data in this type of format requires that the individual acoustic emission events or bursts be characterized. Clearly, this is difficult if the burst rate is high, giving a significant overlap of individual bursts. A considerable amount of acoustic emission data is currently presented in terms of the amplitude distribution of events. The expectation in this approach is that different sources of acoustic emission will produce different distributions of signal amplitude. At least one success has been reported and verified. It has been shown that the acoustic emissions generated by matrix deformation, those generated by matrix-fiber debonding, and those generated by fiber breakage in certain composite materials can be identified and separated by the amplitudes of the measured emissions.[29]

It is also possible to analyze the frequency content of both burst and continuous types of acoustic emission. Relating such measurements to the source mechanism is an extremely complex problem, not only because of the specimen and transducer resonances and frequency dependent absorption effects, but also because of effects caused by the methods of analysis employed. Indeed, some investigators would argue that the information required to distinguish between different acoustic emission sources or to describe the nature of operation of particular sources is simply not available in the frequency content of the acoustic emission signals, particularly using piezoelectric transducers. Other investigators believe that spectral analysis can be of great value and that with proper methods of averaging and smoothing of data, spectral analysis can be used to identify and discriminate acoustic emission sources. Stephens and Pollock[30] were among the first to seriously investigate the use of spectral analysis and give a semiquantitative demonstration of feasibility. Numerous other investigations have been carried out with mixed results. Graham and Alers[31] claim that spectral analysis can be used to distinguish between two separate acoustic emission processes in steel. Woodward and Harris[32] have also claimed success in identifying three sources in the deformation of zirconium, although the results and correlation are not especially clear to the authors. Beattie[33] has written an excellent review paper concerning the digital analysis of acoustic emission signals. He discusses common pitfalls and how different types of analyses of the data might affect the final results.

A recent investigation by Egle *et al.*[34] has provided excellent results and renewed enthusiasm for use of spectral analysis. They have investigated the frequency spectra of the acoustic emission from nodular cast iron tested in compression. Nodular cast iron is an excellent test material, since the acoustic emission from two distinct sources can easily be identified when tested in uniaxial compression.[35] Significant acoustic emission is observed at yield due to dislocation avalanche motion in the ferrite grains, coupled with fracture of scattered pearlite colonies. At higher strains, acoustic emission is observed from the fracture of the graphite nodules. Clearly, if frequency analysis is going to be useful in identifying sources of acoustic emission, one would expect different frequency characteristics from the two sources in nodular cast iron.

The energy spectra were computed from digitized transient acoustic emission signals. The acoustic emission bursts were acquired in groups of 24 events. The Fourier transform of each event was computed, the energy spectrum computed and normalized, and all the spectra in each group of events averaged. A substantial difference was observed in the average frequency content of the emissions associated with dislocation motion and the emissions from fracture of graphite modules. This difference was observed regardless of whether a resonant or flat piezoelectric transducer was used as a detector, however, the choice of sensor did "color" the measured spectra. Large variations were observed in the frequency spectra of individual acoustic emission events so that averaging was required to differentiate the frequency spectra of the sources. Therefore, it would be very difficult to sort acoustic emission events by source in real time if two or more sources were operating simultaneously.

In many instances, the comparison of acoustic emission data from different investigators has been difficult due to the lack of any calibration standards. While this is still a problem, some progress has been made in the past few years. Work at NBS has been progressing toward the development of acoustic emission standards and some very practical standards have evolved. Perhaps the simplest, and yet most efficient, is the breaking of a 0.3-millimeter (mm) Pentel* pencil lead.[36] This has provided a remarkably reproducible simulated acoustic emission source, which is currently used in many laboratories as a calibration source. The use of a pulsed laser to stimulate elastic waves thermoelastically is also receiving attention as a possible source for calibration. This provides an extremely accurate and reproducible source, but at a high cost.

In many instances, if only one measure of acoustic emission is presented, considerable misunderstanding of the data can result. A simple example will suffice. The acoustic emission generated during plastic deformation of uranium-niobium alloys with 5-7 percent** niobium is characterized by large numbers of very small bursts. A system gain of near 110 decibels (dB) is required for detection. Characterization of the acoustic emission activity by ringdown counts or number of bursts would produce large numbers. If this were the only measure of the acoustic emission, the logical conclusion would be that this is a noisy material. Conversly, if the acoustic emission activity were determined by measuring the rms voltage or energy, a small value would be obtained giving the conclusion that this material is very quiet. Figure 1 shows both burst rate

---

* Trademark of Pentel of America Ltd., Torrance, California.

** All compositions given in this paper are weight percent unless otherwise specified.

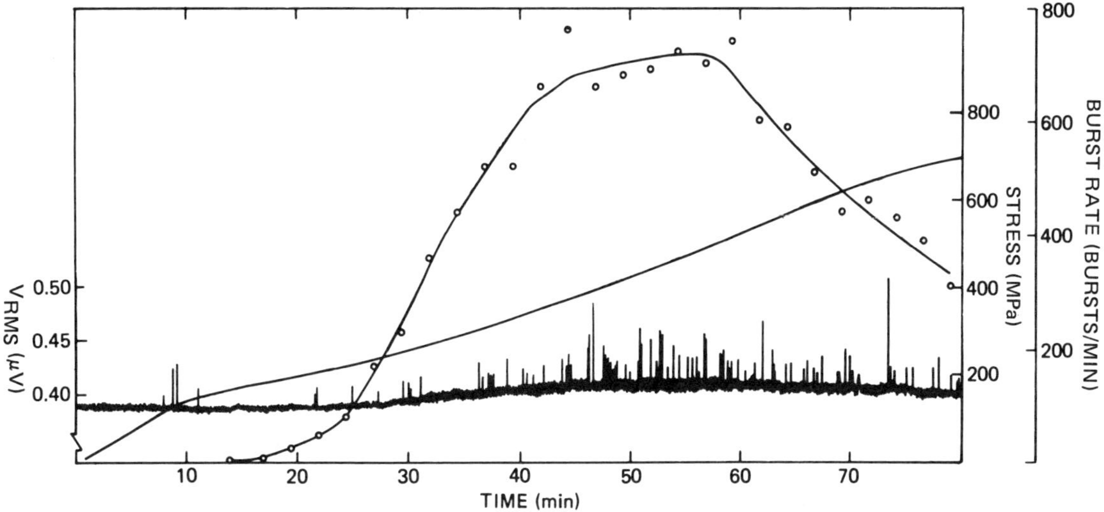

Figure 1. Acoustic emission burst rate and rms voltage (not corrected for noise) referred to the transducer output in tension from U-6 % Nb. The alloy was heated into the γ phase (800°C) and water quenched prior to machining and testing. It was tested at a nominal crosshead speed of 0.05 mm/min, producing a strain rate beyond yield of about 0.0009/min. Total strain during the portion of the test shown was 0.084. Sample gage volume was 0.54 cm^3. For comparison with Figure 2, the maximum burst rate of ~ 700 bursts/min is equal to 1.4×10^6 bursts/(cm^3 · unit strain). (Reference 1)

data and rms voltage for U-6 % Nb alloy tested in tension. A wide disparity of the data in the different formats is clearly evident. Characterization of acoustic emission data will often require the use of more than one measure of acoustic emission activity.

2.4 Concurrent Measurements

A serious drawback in the interpretation of a great deal of acoustic emission data is that acoustic emission is the only microscopically sensitive parameter measured. The identification of the source of the emission under such conditions is difficult at best. In the majority of acoustic emission investigations where the emission is measured during deformation, load-deflection are the only data available to the investigator with which to correlate and interpret his acoustic emission measurements. The macroscopic nature of the load-deflection data makes it impossible to unambiguously identify possible sources of emission. The fact that all of the following dislocation mechanisms have been proposed as the source of acoustic emission at the onset of plastic deformation emphasizes this point. Postulated mechanisms include Luders band formation, slip line or band formation, microcrack formation, dislocation pileup relaxation, twinning, dislocation pinning, dislocation breakaway, the mobile dislocation density, slip advances, and rapid source operation i.e., dislocation multiplication.

Published work[35,37] as well as data to be presented later in this paper, demonstrate the usefulness of measuring a microscopically sensitive experimental parameter in addition to acoustic emission. In these studies, dislocation damping (internal friction) was measured, as well as acoustic emission. The dependence of dislocation damping in the kilohertz frequency range on dislocation density and average dislocation loop length is well known and can be used to help identify particular sources of emission. Other investigators have employed simultaneous measurements of high frequency (megahertz) attenuation with their acoustic emission measurements.[38,39] In all internal friction and attenuation measurements, an elastic wave is introduced to drive the sample, usually by some type of driving transducer. Damping is a measure of the decay or energy dissipation of the driving wave. Care must be taken that the driving wave is not measured as acoustic emission. An experimental dilemma arises in that in order to avoid measuring the driving wave, the system gain may be set so low that the acoustic emission produced during deformation is not detected.

An investigation in which both the acoustic emission and electrical resistivity were measured as a function of crack opening displacement leaves little doubt that the acoustic emission was generated during crack growth.[40] Excellent correlation between the acoustic emission and changes in the electrical resistivity were obtained. Numerous other parameters are available which might give added insight in the analysis of acoustic emission data. Possible parameters would include dislocation density, molulus defect, microstructural changes, surface characteristics, and even accurate measurements of the plastic strain and plastic strain rate.

The current controversy over the role of inclusions (particularly MnS-type inclusions in steels) is another illustration of the necessity of measurements in addition to acoustic emission in order to determine the nature of the acoustic emission source. There have been a number of careful investigations[41-43] of the acoustic emission from rolled steel plate containing elongated MnS inclusions. The acoustic emission from tensile samples with their tensile axis perpendicular to the rolling plane (short transverse direction) is considerably different from the acoustic emission from samples with their tensile axis in the rolling plane. The major differences are that significant acoustic emission continues well beyond yield in the short transverse samples, and the amplitude distribution of the acoustic emission differs for the two testing directions. The additional acoustic emission from the short transverse samples has been attributed to debonding of the MnS-type inclusions from the steel matrix and/or fracture of the inclusions.

Jaffrey[44] has recently investigated the behavior of manganese sulfide inclusions during deformation in a resulfurized plain carbon-manganese steel. He observed that MnS stringers tested parallel to their long axis neither debonded from the steel matrix nor fractured during uniaxial tensile deformation until plastic strains exceeded at least 26 percent. Spherical MnS inclusions never fractured during tensile deformation and only debonded during advanced necking just before fracture of the sample. Neither decohesion nor fracture of the sulfide inclusions occurred during uniaxial compression at strains up to 30 percent.

The steel investigated by Jaffrey contained relatively high purity MnS inclusions. Manganese sulfide-type inclusions in commercial structural steels are more complicated. They contain other phases and additional elements. Thus, as pointed out by Jaffrey, manganese sulfide-type inclusions in commercial steels may not deform or adhere to the matrix as well as the pure sulfides. Furthermore, the samples were tested parallel to the plane of the stringers rather than perpendicular to it. Other possibilities are that the acoustic emission associated with MnS inclusions may not arise from fracture or decohesion of MnS, but rather from failure of other inclusions or cementite, or from localized dislocation motion in the matrix near the inclusons.[45] The local flow arises from the stress concentrations near the inclusions caused by the different elastic and plastic properties of the inclusions and the matrix.

McBride and coworkers[46] have provided an excellent example of the correlation of acoustic emission with microstructural observations in the fracture of 7075 aluminum. They found quantitative agreement during slow crack growth between the amplitude distribution of burst acoustic emissions and the area size distribution of intermetallic inclusions as measured on the polished fracture face. No emissions were detected during crack propagation either when inclusions were not present (high purity samples) or when the alloy was in the "0" condition. The matrix in the "0" condition was so soft that inclusions did not fracture.

2.5 Testing Variables

The proper choice of testing variables can be of great value in attempting to isolate and identify particular sources of emission. In many situations, it is possible to select testing conditions such that a particular source of emissions will act independently during some portion of the test cycle. If emission is then measured during that portion of the test, it can be directly related to the operating source. Several investigations have demonstrated that the simple comparison of acoustic emission measurements carried out in uniaxial tension and compression can be used to investigate the fracture of precipitates[47–49] and the breakup of graphite nodules in cast iron.[35] These investigations have made use of the fact that brittle particles in a matrix will fracture at a significantly lower level of strain in tension than in compression. The majority of the acoustic emission from 7075-T6 aluminum (and essentially all of it from 7075-T651 aluminum) is characterized by a broad maximum in emission versus strain centered at about 1.7 percent plastic strain. This peak had been widely attributed to dislocation processes, but has now been shown to arise from precipitate fracture through the use of both tension and compression[47–49] measurements supported by simultaneous internal friction measurements.[47] In some systems, it is possible to activate different deformation mechanisms using different testing modes. For example, Figure 2 shows acoustic emission data for U-2.2 % Nb tested in both compression and tension. The difference in burst rate is striking. Substantial twinning is believed to occur in tension, while in compression only slip occurs. Similar differences between tension and compression are observed for a number of pure hcp metals.

Selecting the strain rate can also be important. Serrated or

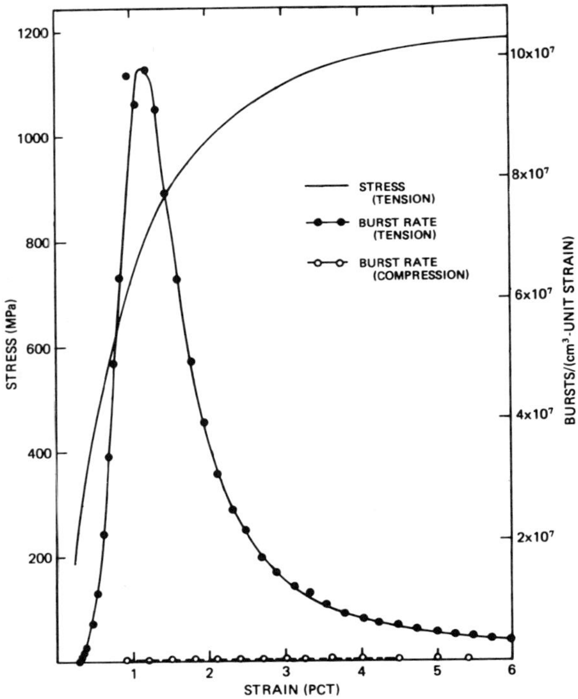

Figure 2. Acoustic emission burst rate versus strain in tension and compression from U-2.2 wt. % Nb. The alloy was heated into the γ phase (900°C) and furnace cooled prior to testing. The cooling rate in the $\gamma \rightarrow \alpha + \gamma_2$ phase transformation temperature range near 650°C was ~810°C/hr. The alloy was tested at a nominal crosshead speed of 0.05 mm/min. Sample gage volumes were 0.46 cm^3 (tension) and 0.89 cm^3 (compression). (Reference 1).

Portevin-LeChatelier yielding only occurs at a given temperature within a limited range of strain rates. Deformation of zinc at slow strain rates is reported[50] to be due to slip, however, at high strain rates the deformation is basically due to twinning. While this certainly does not give a complete overview, these examples do give an indication of the importance of testing parameters.

2.6 Artifacts

Caution must be exercised in an acoustic emission investigation with respect to the operational noises inherent in any deformation testing machine. Results have been published warning of possible noise during testing from such sources as hydraulic valves, clutch slippage, and pin and grip noise.[17] In most cases, noise can be eliminated through the use of good experimental techniques, and/or by application of the Kaiser effect. Low level, subtle machine noise might not be a factor in testing certain materials and yet be critical in the testing of others. For example, the testing of beryllium during deformation produces easily detectable emissions at a system gain of 85 dB.[51] However, measurement of the acoustic emission generated during the deformation of austenitic stainless steel at low strain rates requires a system gain of 110 dB, making the existence of machine noise critical. In certain cases the linearity of crosshead motion of the testing machine is as important as operational noise. Figure 3a shows acoustic emission data for pure aluminum (99.9995% Al) tested with an older Instron machine. The data are similar whether the material is tested in tension or compression. While the data are certainly striking, they are an artifact resulting from the testing machine. The crosshead speed of the testing machine was found not to be constant. The actual crosshead motion is shown in Figure 3b. The average rates were found to be 3.3×10^{-4} in./min. for the slower rate and 7.5×10^{-3} in./min. for the faster rate. The maxima in the emission occur when the crosshead is moving at the faster rate, as shown in Figure 3b. Similar effects have not been observed in other materials tested on the same machine. The normal proportionality between rms emission and the square root of the strain rate was observed in the high purity aluminum when the nominal crosshead velocity was

3(a)

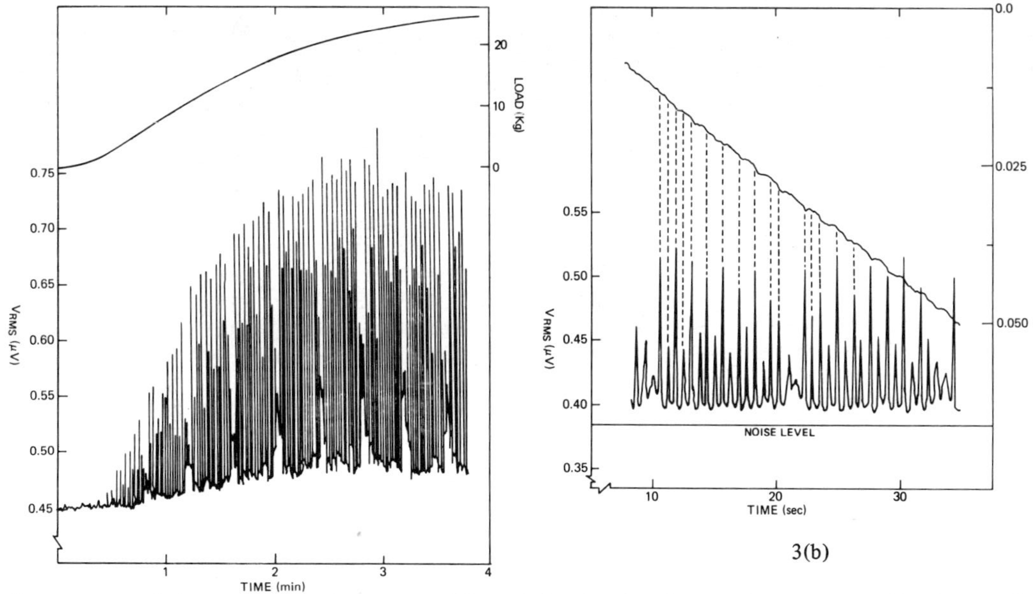

3(b)

Figure 3. Acoustic emission rms voltage (not corrected for noise) referred to the transducer output and load versus time from 99.9995% aluminum in tension. The metal was tested at a nominal crosshead speed of 0.05 mm/min, producing a strain rate beyond yield of about 0.005/min. Sample gage volume was 0.26 cm^3. (a) Behavior near yield. Total strain during the portion of the test shown was 0.018. (b) Acoustic emission and actual crosshead displacement versus time. (Reference 1).

changed. The origin of the particular sensitivity of acoustic emission from high-purity aluminum to nonlinearity of the crosshead velocity is not known.

3. MODELS FOR ACOUSTIC EMISSION FROM DISLOCATION SOURCES

A number of potential mechanisms exist whereby the motion of a dislocation can create elastic waves, including relaxation of the elastic stress field in the lattice caused by the passage of a dislocation, radiation from acceleration of a dislocation, and annihilation of a dislocation. The stress field relaxation appears to be the largest effect. These effects will be discussed in more detail subsequently in this section.

3.1 Surface Displacement From A Moving Dislocation

One approach to calculation the expected magnitude of the displacement of the surface from the motion of a dislocation within the material is to treat the stress field of a dislocation loop as a sum of force dipoles buried in the material. Sinclair[52] has written, from more general expressions given by others, analytic expressions for the dynamic elastic Green's tensor for a half-space at the epicenter location (the point on the surface intersected by a normal to the surface passing through the source). The Green's tensor in this application relates the displacement at any point in the elastic body to a point force acting at the source location.

Scruby, Wadley, and Sinclair[53] calculated the peak amplitude of the surface displacement from the first (longitudinal) wave arrival at the epicenter caused by the growth of a dislocation loop in an isotropic material. The loop was assumed to lie a depth D below the surface at 45° to the tensile axis, and to grow at constant velocity, v, from zero radius to the final radius r. The peak displacement amplitude, U, is then calculated to be

$$U = \frac{brvC_2^2}{D\,C_1^3} \tag{6}$$

where b is the Burger's vector, C_1 is the longitudinal wave speed, and C_2 is the shear wave speed. For n multiple loops expanding cooperatively to the same final radius, the peak displacement amplitude is given by

$$U_n = \frac{nbrvC_2^2}{D\,C_1^3}\,. \tag{7}$$

Several features of these expressions should be noted. First, the magnitude of the surface displacement is directly proportional to how far the dislocations move (the final loop radius) and their velocity of motion. The amplitude is also material dependent through the shear and longitudinal wave velocities. Using typical values for aluminum (C_1 = 6400 m/sec, C_2 = 3200 m/sec, b $\approx 2.9 \times 10^{-10}$ m), if the dislocation velocity is 200 m/sec and the source is 4 centimeters from the epicenter, the calculated peak displacement for one dislocation loop expanding to the radius of a single crystal specimen (taken to have a 4 millimeter diameter) would be 1×10^{-13} m. This displacement is probably detectable by piezoelectric transducers. If, however, the same source were in a polycrystal with a 40 micrometer (μm) grain size, then the maximum loop radius would be 20 μm, and the calculated displacement would be 1×10^{-15}, which is probably not detectable. Simmons and Clough[54] have also calculated the displacement from a dislocation loop expanding at constant radial velocity using a Green's tensor approach. Eq. (6) can be derived[159] from their result, except for a factor of two which arises because Simmons and Clough consider displacement in an infinite space and Scruby *et al*. consider displacement at the surface of a half-space.

Thus, these calculations predict that in commercial materials where the glide distance of a dislocation is unlikely to exceed

20-50μm, the motion of a single dislocation is unlikely to produce elastic waves detectable by currently available transducers which can detect displacements of order 10^{-14} m. (See Section 2.2)

This conclusion is in accord with a substantial body of experimental evidence from which it is clear that only a small fraction of the dislocation motion produced by deforming polycrystalline metals creates elastic waves large enough to be detected as acoustic emission.

For example, the plastic strain rate $\dot{\epsilon}_p$ is given by

$$\dot{\epsilon}_p = \lambda \, b \, \rho_m \, \overline{v}, \tag{8}$$

where λ is a geometric constant, b is the Burger's vector, ρ_m is the density of moving dislocations, and $\overline{v}$ is their average velocity. Since, in a typical constant crosshead velocity tensile or compression test, the plastic strain rate at large strains exceeds the plastic strain rate near the onset of plastic flow, the amount of dislocation motion increases beyond yield. However, the acoustic emission (in the absence of other sources such as precipitate fracture, strain-induced phase transformations, cracking, twinning, etc.) generally drops to low values (or zero) at large strains. If the motion of all dislocation were being detected, an increase in the acoustic emission from the onset of plastic flow to at least the ultimate tensile strength would be expected. Gillis and Hamstad,[55] analyzing data generated by Fisher and Lally,[56] estimated the fraction of dislocation motion producing acoustic emission to be approximately 1%. Using a simple energy calibration scheme, one of the authors[57] has estimated the maximum ratio of the energy released by the events responsible for acoustic emission to the total energy dissipated in plastic flow of beryllium to be approximately 0.7 percent, in essential agreement with the estimate of Gillis and Hamstad.[55] Even this modest fraction was only achieved in an 0.4 percent strain range near the initial yield point load drop in a particularly noisy sample of beryllium.

Therefore, with the possible exception of single crystals or large grained pure samples under favorable circumstances, elastic waves large enough to be detected as acoustic emission can be produced from dislocation motion only if many dislocations move nearly simultaneously within a small volume of material. Furthermore, the packet of dislocations must move far enough and fast enough, or its motion will remain undetected. Since these conditions are not often simultaneously satisfied, only a small fraction of the dislocation motion associated with plastic flow produces acoustic emission.

3.2 Mechanisms for Coordinated Dislocation Motion

Various microstructural and material variables will affect the probability for simultaneous dislocation motion, the glide distance, and the velocity of motion. The various models for the production of acoustic emission by dislocation motion have generally concentrated on only one of these aspects. James and Carpenter[58] proposed a mechanism whereby the simultaneous motion of many dislocations within a small volume of material could occur. It is well known that dislocations can be pinned by a variety of mechanisms. James and Carpenter proposed that as the stress is increased on a set of pinned dislocations, one or a small number of the dislocations break away from their pins. The small stress wave produced by the breakaway can cause the unpinning of additional nearby dislocations and result in an avalanche of dislocation motion within a small volume of material over a short time interval. In their experiments with LiF, James and Carpenter estimated the total length of dislocation line involved in an average burst to be $\approx 2{,}000$ cm.

Agarwal, Frederick, and Felbeck[16] proposed a different source mechanism. They envisioned that as a sample is loaded, a stress is eventually reached at which Frank-Read or grain boundary sources are activated. The newly created dislocations move along the slip plane until they are stopped by an obstacle, such as a grain boundary, precipitate, or other dislocations. The source, once activated, continues to rapidly produce dislocations until the back stress arising from the pileup of the new dislocations against the obstacle becomes large enough to shut off the source. Agarwal *et al.* calculated the axial displacement in the sample from this process and assumed that this axial displacement is what is detected by the transducer as acoustic emission. (This assumption is a simplification, since it is the elastic waves released by the dislocation motion which are actually detected.) They also calculated the minimum displacement detectable by their piezoelectric transducer to be approximately 7×10^{-15} m. Assuming a slipped region diameter of 10 μm and 13 dislocations involved, the minimum detectable event for their sample geometry and 2024 aluminum corresponds to the creation and motion of approximately 10^{-2} cm of dislocation line.

In a series of papers, Kuribayashi, Kishi, and others[59–61] have examined the release of the stress concentration caused by the pileup of dislocations at a grain boundary as a source of acoustic emission. Initial deformation proceeds inhomogeneously, with the most favorably oriented grains yielding first. Dislocations in these yielding grains pile up at their grain boundaries, producing a stress concentration in neighboring, unyielded grains. As the stress is increased, either the pileup breaks through the boundary or grain boundary sources are activated. A packet of dislocations then moves across the previously unyielded grain, essentially as envisioned by Agarwal *et al.*,[16] releasing the stress concentration caused by the pileup. Kuribayashi *et al.* calculated the motion of the dislocations in the stress field (the applied stress plus the stress concentration from the pileup) and from that the time required for the dislocations to cross the grain. This time will, of course, depend on the grain size, and will decrease with decreasing grain size. They then propose that the passage of dislocations through the (assumed spherical) yielding grain sets it into oscillation, and calculate the spectral distribution of the energy released by this oscillation. The calculated spectral distribution is an oscillatory function of grain size. In addition, the measurement of acoustic emission is often made with a resonant transducer on a signal bandpassed for a relatively small band of frequencies. Their calculations predict that for a given narrow measurement frequency band, the observed acoustic emission will be a maximum for some grain size (~20 μm at 140 KHz and 50-100 μm at 30 KHz for aluminum), decreasing for larger and smaller grained samples. The grain size at which the predicted maximum in detected emission occurs decreases with higher measurement frequencies. Kuribayashi *et al.*[60] present rms measurements on the acoustic emission from 99.9 percent pure aluminm as a function of grain size in agreement with the above predictions; however, measurements by other authors yield different results. Observations on the effect of grain size on acoustic emission will be discussed further in Section 4.9 (Grain Size). The mechanism for release of stress concentrations proposed in this model appears reasonable, however, it is unclear how the postulated grain oscillation can occur. An alternative explanation proposed by Fleischmann and coworkers[62] for certain observed frequency effects is described below.

3.3 Frequency Effects

Fleischmann *et al.*[62] have considered the effect of the velocity and glide distance of a moving dislocation on the frequency spectrum of the ultrasonic waves produced. The dislocation is assumed to move at constant velocity v from the time motion begins until it is stopped by an obstacle after it has travelled a distance d. Furthermore, the moving dislocation is treated as an ultrasonic source which produces a pure step displacement perpendicular to its plane of motion. The frequency spectrum of the wave created by such a source has a maximum (to first order) at a frequency

$$f = \frac{v}{2d} \quad \text{(f in hertz).} \tag{9}$$

If the glide distance is taken to be the grain diameter, then the frequency at which the maximum acoustic emission energy is produced increases for decreasing grain size. Furthermore, they observed[62, 63] experimentally that the frequency of the maximum shifted to higher frequencies as deformation progressed. This shift was attributed primarily to a reduction in the glide distance of the dislocations with increasing deformation. For 99.99 percent aluminum with 1,000 μm grain size after 1 percent plastic deformation, the time required for dislocations to move between obstacles was calculated from their observations to be about 0.35 microseconds. The analysis is presented in terms of a single moving dislocation, however, it is unchanged if a packet of moving dislocations is considered instead.

3.4 Solute Drag

Kuribayashi and Kishi[64] have also proposed that the elastic waves detected as acoustic emission do not originate from the elastic stress relaxation due to dislocation glide, but rather arise from the energy radiated by accelerating dislocations. Energy radiated by an accelerating dislocation is analogous to Bremsstrahlung radiation from acceleration of a charged particle. They proposed that if the lattice contains solute atoms mobile at the test temperature, then these solute atoms would produce a viscous type of drag on moving dislocations. The viscous drag would result in more uniform motion of the dislocations, with dimished acceleration or deceleration and, therefore, reduced acoustic emission. They observed a reduction in the height of the acoustic emission rms peak near yield in aluminum single crystals with increasing magnesium content. An alternative explanation is that the solute atoms reduce the velocity of the packet of moving dislocations and thereby reduce the size of the elastic wave produced in accordance with equation 7.

3.5 Acoustic Emission Versus Strain

A number of attempts have been made to explain the shape of the curve of the amount of acoustic emission observed as a function of strain. Several investigators have noted that the shape of this curve could be duplicated (for some materials) by an expression for the mobile dislocation density ρ_m, as a function of plastic strain ϵ_p, proposed by Gilman[65] to be

$$\rho_m(\epsilon_p) = [\rho_m(o) + M\,\epsilon_p]\; e^{-\Phi\epsilon_p} \tag{10}$$

where M and Φ are constants. Unfortunately, this expression was derived to represent the mobile dislocation density as a function of strain for a constant stress creep test. In this kind of test, the dislocations activated by the application of a constant stress move and are eventually annihilited or blocked by obstacles, resulting in an asymptotic decrease in the plastic strain rate toward zero. If the constants M and Φ are adjusted to provide the desired proportionality between the mobile dislocation density and the measured amount of acoustic emission for a typical constant crosshead velocity tensile test, then Eq. (9) predicts that the mobile dislocation density falls rapidly toward zero shortly after yield. Since the plastic strain rate does not decrease, but instead continues to increase somewhat following yield in a constant crosshead velocity test, the diminishing number of mobile dislocations with increasing strain predicted from Eq. (9) would have to move farther and farther at higher and higher velocities in order to maintain the plastic strain rate. This is at variance with the entire body of experimental evidence on the behavior of dislocations with increasing strain following yield.

Teleman[66] provided an alternative explanation for the shape of the acoustic emission versus strain using an entirely different premise. He proposed that the density of sources capable of producing detectable avalanches or packets of moving dislocations had a distribution in strain, specifically a Weibull distribution. Thus, the number of sources which have operated up to a strain ϵ is given by $N_s(\epsilon)$

$$N_s(\epsilon) = g[1 - e^{-d(\epsilon - \epsilon_0)^f}] \quad (\epsilon > \epsilon_0) \qquad (11)$$

where ϵ_o is the strain necessary to operate the most easily operated source, g is the total number of sources, and f and d are constants. The values of f and d determine the shape of the distribution. For d=1 and f near 3.6, the distribution is similar in shape to a normal distribution. For d=1 and f<3.6, the Weibull density function is skewed to higher strains. The asymmetry is such that more sources operate at strains above the strain corresponding to the maximum in the density function than below it. A discussion of the Weibull distribution is presented in Reference 67.

The rate at which sources would operate as a function of strain would then be given by

$$\frac{d\, N_s\,(\epsilon)}{d\,\epsilon} = g\,\frac{e^{-d(\epsilon - \epsilon_0)^f}}{(\epsilon - \epsilon_o)^{1-f}}\, f\, d \quad (\epsilon > \epsilon_o). \qquad (12)$$

This approach assumes that once a source is used it cannot be used again, which is appropriate when the Kaiser effect is observed. The sources could instead be distributed as a function of stress rather than strain, in which case the stress σ would be substituted for strain in Eqs. (11) and (12). If the events were distributed in stress, then recasting the cumulative distribution function N_s and event rate as a function of strain requires the functional relation between stress and strain. The authors are not aware of any efforts to apply this equation to describe the dependence of acoustic emission on strain (or stress), nor to develop a physical interpretation for the constants d and f. The Weibull distribution, however, has been used to describe the amplitude distribution of acoustic emission signals.[41,43]

A third qualitative explanation for the dependence of

acoustic emission on strain is that, as deformation proceeds, more and more obstacles to the glide of the dislocation packets are produced. These obstacles, such as forest dislocations and dislocation tangles, progressively reduce the distance the dislocation packets move and thereby, in accord with Eq. (7), reduce the size of the elastic waves produced until most or all of them are undetectable. This process would imply a shift in the distribution of signal amplitudes toward smaller amplitudes with increasing strain through the acoustic emission peak at the onset of plastic flow. We are not aware of evidence for such a shift in amplitude distribution, however, it is difficult to measure the actual signal amplitude distribution for metals when the acoustic emission is clearly from dislocation motion because the signal is usually of the continuous type, where the observed waveform is the superposition of many small bursts.

Finally, the shift observed and predicted by Fleischmann *et al.*[62] in the frequency spectrum, of the energy released as acoustic emission, to higher frequencies with increasing strain also affects the shape of the observed acoustic emission versus strain curve. Since most measurements are made over a narrow frequency band, the shift in frequency with strain means that the fraction of the total energy released which falls within the measuring bandpass changes with strain.

3.6 Experimental Evidence

There is a substantial body of experimental data and quantitative calculations available (some of which have been discussed above) which can be used to test the validity of the concepts for creation of acoustic emission by dislocation motion. As indicated previously, the nearly simultaneous motion of a number of dislocations within a small volume of material appears to be normally required in order to produce elastic waves large enough to be detected as acoustic emission. Certain microscopic metallurgical processes would appear to primarily affect the probability of the formation of a packet or avalanche of moving dislocations. The pinning of dislocations by relatively weak pins, so that the dislocations can break away from their pins as the stress is raised, is one such process. The original experiments of James and Carpenter[58] in LiF, on which their model was based, demonstrated that the acoustic emission event rate was not proportional to the total dislocation density, or the mobile dislocation density, but to the rate of change of the mobile dislocation density. This implies that a process associated with the conversion of immobile to mobile dislocations, rather than the creation of new dislocations, was responsible for acoustic emission in their samples. They proposed an avalanche-type dislocation breakaway from pins as the acoustic emission source mechanism. There is considerable additional evidence consistent with this view, particularly in circumstances where dislocation breakaway is known from other experiments to occur. Load drops at yield can result from the breakaway of dislocations from solute pinning atmospheres. Hence, relatively high levels of acoustic emission should be expected, and are, in fact, often observed in metals when a load drop occurs at yield because the probability of formation of dislocation avalanches is relatively high. The acoustic emission from iron[37] (bcc, pinning from carbon interstitials) and 5083 aluminum[68] (fcc, pinning from substitutional Mg) is shown in Figures 4 and 5. Acoustic emission behavior like that shown in Figure 5 for 5083 aluminum has been reported[69] for 5086 aluminum, which is an aluminum-magnesium alloy very similar to 5083 aluminum. The load drop is believed to be from dislocation breakaway in these materials.

Serrated or Portevin-LeChatelier yielding has been shown to result at least in part from repeated locking and unlocking of dislocations by solute atoms during deformation. This mechanism would also be expected to increase the number of dislocation avalanches. Acoustic emission has been observed with both Type A serrations (the large load drops associated with initiation of Portevin-LeChatelier deformation bands)

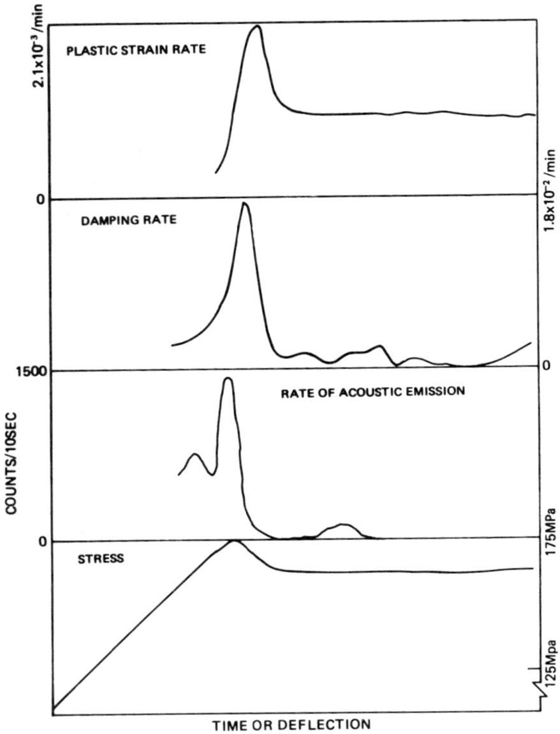

Figure 4. Acoustic emission count rate, time derivative of the amplitude independent damping, stress, and plastic strain rate for pure Armco iron versus time in tension. The metal was tested at a nominal cross-head speed of 0.05 mm/min, producing a strain rate beyond the load drop of 0.001/min. Sample gage volume was 0.2 cm^3. (Reference 37)

and Type B serrations (small load drops associated with unlocking of the deformation band as it travels the length of the specimen) in a number of materials.[53, 70–78] An example of the acoustic emission associated with both types of serrations is shown in Figures 6 and 7 for 5083 aluminum.[68] Hartman[71] has correlated the energy released as acoustic emission in

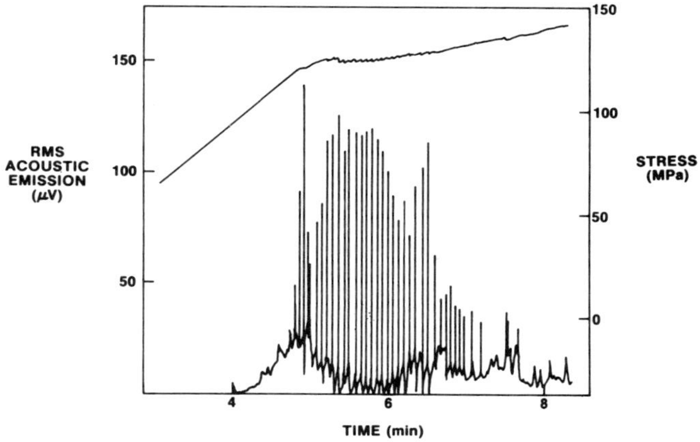

Figure 5. Acoustic emission rms voltage (not corrected for noise and referred to the transducer output) and stress versus strain for 5083-0 aluminum in tension. The alloy was tested at a nominal cross-head speed of 0.05 mm/min, producing a strain rate of about 0.0023/min in the region of initial yielding shown. Total plastic strain in the portion of the test shown as about 0.006. Sample gage volume was 0.26 cm^3. (Reference 68)

deadweight loaded α-brass at room temperature with the stored energy between strain increments. Takagaki and Mukherjee[78] have also investigated acoustic emission during serrated yielding of α-brass and concluded that sudden dislocation motion both from unpinning and dislocation multiplication contribute to the acoustic emission.

Another mechanism has also been proposed for the origin of load drops at yield (or yield points). Hahn[79] proposed that the activation of dislocation sources – most likely grain boundary sources – under conditions where the stress to generate a dislocation exceeds the stress necessary to move it would produce a yield point. Such a process is likely to form groups of rapidly moving dislocations, and thereby contribute to acoustic emission. It is similar to the mechanism proposed

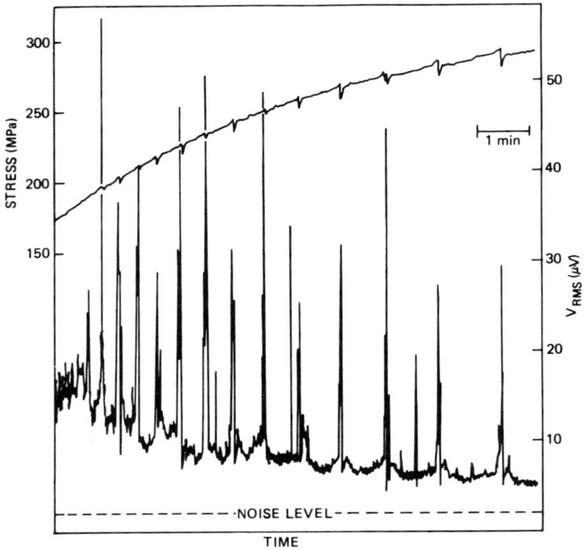

Figure 6. Acoustic emission rms voltage (referred to transducer output) and stress versus time for 5083-0 aluminum in tension in a portion of the deformation exhibiting Type A Portevin-LeChatelier yielding. The sample was tested at a nominal crosshead speed of 0.25 mm/min, producing a strain rate of about 0.012/min in the portion of the test shown. The figure includes plastic strains from 0.015 to 0.13. Sample gage volume was 0.26 cm^3. (Reference 68)

by Agarwal *et al.*.[16] The load drop at yield in beryllium is believed[80] to be from the generation of new dislocations rather than from breakaway of existing dislocations from pinning points.

The breakaway and generation processes may be distinguishable experimentally by measuring the internal friction (damping) as a function of strain near yield. The amplitude-independent dislocation damping depends on the fourth power of the average dislocation loop length between pinning points and linearly on the dislocation density.[81, 82] Thus, a breakaway

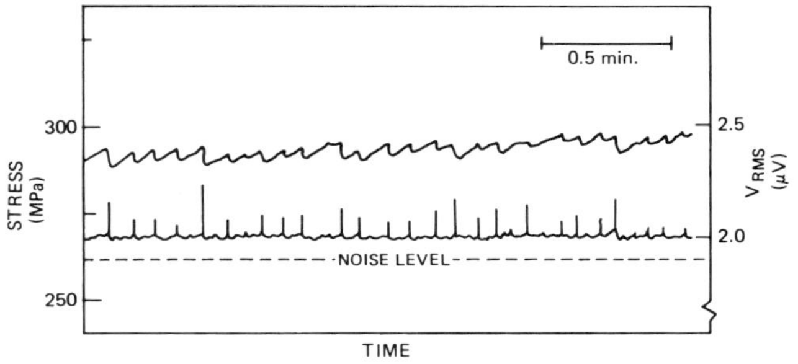

Figure 7. Acoustic emission rms voltage (referred to transducer output) and stress versus time for solutionized-and-quenched 5083 aluminum in a portion of the deformation exhibiting Type B Portevin-LeChatelier yielding. The sample was tested at a nominal crosshead speed of 0.05 mm/min, producing a strain rate of about 0.0028/min in the portion of the test shown. The figure includes plastic strains from 0.065 to 0.072. Sample gage volume was 0.11 cm^3. (Reference 68)

process which considerably increase the average loop length should cause a significantly greater increase in the damping than would be observed from a dislocation generation process which primarily increases the dislocation density. The dislocation damping for iron exhibits a substantial peak at the yield point, as shown in Figure 4. On the other hand, the change in damping at yield in beryllium is small (Figure 8), which is consistent with the yield point mechanism in beryllium being the generation of new dislocations, rather than a breakaway process. Beryllium also exhibits a second acoustic emission peak, centered at about 1% plastic strain, which has been attributed to dislocation motion.[51] The increase in damping in the vicinity of the second acoustic emission peak is much larger than the increase associated with the acoustic emission peak at yield, which is consistent with the second peak arising from a dislocation breakaway process.[83]

The magnitude of the acoustic emission peak associated with the load drop at yield should be sensitive to heat treatment,

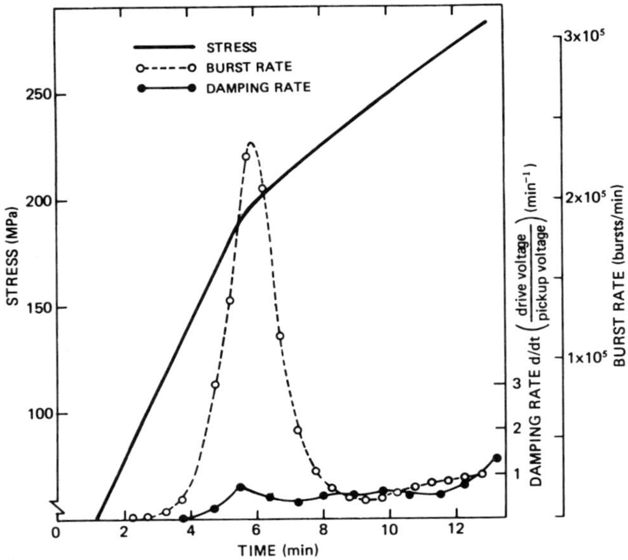

Figure 8. Acoustic emission burst rate, time derivative of the amplitude independent damping, and stress versus time from vacuum hot pressed beryllium in tension. Details of the beryllium composition and processing are given in Reference 51. The acoustic emission and damping were measured on different samples. The acoustic emission sample was tested at a nominal crosshead speed of 0.05 mm/min, producing a strain rate beyond yield of about 0.0016/min. Sample gage volume was 0.38 cm^3.

temperature, previous mechanical working, concentration of the pinning points (solute atoms or precipitates), and other factors which influence the extent and strength of the dislocation locking. The magnitude and existence of yield points are sensitive to these factors. Several of these effects were observed in beryllium[51] where changes in heat treatment and purity altered or eliminated the load drop at yield and produced very large changes in the amount of acoustic emission detected near yield. It should be noted that the amount of emission is not necessarily correlated with the size of the load drop at yield because metallurgical factors which increase the likelihood of dislocation avalanches may also change the speed or distance the avalanches move. Nakasa[84, 85] observed a three times reduction in count rate for vanadium – 600 ppm boron at the yield point after neutron irradiation, but the load drop at yield, if anything, increased.

Precipitation in age-hardening systems is another phenomenon which can affect the probability of dislocation avalanche formation, the glide distance of the avalanches, and possibly their velocity of motion. In their original paper, Agarwal *et al.*[16] proposed that age-hardening precipitates would always decrease acoustic emission from dislocation motion because the precipitates reduce both the size of Frank-Read sources and the glide distance of the dislocations. They observed results consistent with this view in 2024 aluminum, and results in other alloy systems have also been interpreted on this basis. However, Heiple *et al.*[86] observed a substantial increase in acoustic emission at the onset of plastic flow in some age-hardening stainless steels for short aging times which produced small, cuttable precipitates. They interpreted this result, combined with results from aluminum and other alloys, to indicate that age-hardening precipitates alter acoustic emission primarily by affecting the probability of dislocation avalanche formation. Scruby *et al.*[14] have also observed an increase in acoustic emission in certain aluminum alloys after particular aging conditions. They proposed that the increase was associated with strain localization, promoted by cuttable precipitates. The proposed mechanisms and associated experimental evidence for the effect of precipitates on acoustic emission from dislocation motion are described in Section 4.8 (Heat Treatment).

It has been argued[87] that dislocation breakaway cannot account for the observed acoustic emission, since the maximum plastic strain attainable by movement of an average of 1/2 grain diameter by dislocations present in a typical recrystallized sample is approximately 0.15 percent. New dislocations must therefore be generated early in the deformation process. However, as discussed previously, only a small fraction (less than 1 percent) of the total dislocation motion within a deforming sample contributes to the observed acoustic emission. Thus, there is enough dislocation motion present in a material which does not contribute to the acoustic emission to account for the plastic deformation and still have breakaway of groups of pinned dislocations be the major source of acoustic emission.

3.7 Additional Quantitative Descriptions of Dislocation Sources

There has been a considerable effort and substantial progress over the last few years in obtaining quantitative descriptions of source events from the measured acoustic emission signals. This work has been carried out primarily by groups at the U.S. National Bureau of Standards, Cornell University, and the U.K. Atomic Energy Research Establishment (Harwell). The development of capacitive transducers and laser transducers has made possible very broadband, absolute measurements of surface displacements. Spectacular agreement between calculated and measured surface displacements has been achieved with simulated acoustic emission sources such as the fracture of a glass capillary or a pencil lead.[88–90] Quantitative results have also been obtained at Harwell[11–14] from real acoustic emission events, using a specially shaped specimen called a Yobell. Work in this area has been recently reviewed extensively by Lord[2] and will not be discussed in detail here.

As discussed in Section 2.2, the broadband transducers which can make absolute, quantitative measurements of surface displacements are unfortunately not as sensitive as piezoelectric transducers commonly used for acoustic emission measurements. Piezoelectric transducers are two orders of magnitude more sensitive. For this reason, quantitative measurements of surface displacements attributed to dislocation motion have only been reported once.[14] Calculations of the surface displacement predicted from dislocation motion have been made, as indicated previously. The mathematics involved in these calculations is formidable, but the complexity should not be allowed to obscure the fact that a number of important approximations are involved and that the detailed mechanism(s) by which moving dislocations create elastic waves are imperfectly known. Nevertheless, significant progress is being made in these quantitative calculations which shed valuable light on the functional dependence of the acoustic emission on parameters of dislocation motion and on the size of the acoustic emission signal expected from dislocation motion.

In addition to the work described in the preceding sections, there have been a number of other calculations and measurements relevant to the characteristics of elastic waves generated by dislocation motion. Malen and Bolin[91] estimated the stress amplitude of an elastic wave generated by the sudden homogeneous change in inelastic distortion $\Delta\epsilon^*$ in a small volume v^*. Using a piezoelectric transducer sensitivity, they calculated the product of $\Delta\epsilon^*$ and v^* required to produce a 10 μv signal from the transducer to be 10^{-16} m^3. If the distortion is produced by homogeneous slip within v^*, then $\Delta\epsilon^* v^*$ is given by

$$\Delta\epsilon^* v^* \approx b\, \rho_m\, \Delta\bar{x}\, v^* \tag{13}$$

where b is the Burger's vector, $\Delta\bar{x}$ is the average distance the dislocations have moved, and ρ_m is the density of moving dislocations. Applying these relations to the formation of slip bands observed by James and Carpaneter in LiF, Malen and Bolin calculated that the acoustic emission signals should have had an amplitude of 3,000 μv rather than the 20 μv which was actually observed. Thus, less than 1 percent of the dislocations forming the slip band contribute to the acoustic emission. This result is consistent with other observations. Tandon and Tangri[92] observed in 35 μm grain size silicon iron that approximately 400 times as many slip bands were observed as acoustic emission events were detected for deformation prior to the upper yield point. Day[93] even observed torsional strain bursts in zinc resulting in displacements of 100A without any detectable acoustic emission.

Using an approach similar to Malen and Bolin, Ono[5] and Hsu and Ono[74] have derived an expression for the rms output of a narrow band resonant transducer (Vrms) excited by numerous small signals random in time ("continuous" emission). For moving dislocations as the source, the resulting expression is

$$V_{rms} = K\, v \sqrt{\rho_m \ell} \tag{14}$$

where K is a constant, v is the average glide speed between obstacles of the moving dislocations, ρ_m is the mobile dislocation density, and ℓ is the average length of moving dislocation segments. As does Eq. (7), this expression predicts that the acoustic emission increases with increased dislocation velocity, increased numbers of moving dislocations, and increased length of moving dislocations.

Telelman[66] calculated the energy ΔE_g released by yielding one cubical grain of side d by moving n dislocations across the grain with a driving stress τ. His expression was

$$\Delta E_g \simeq n\, b\, \tau\, d^2 . \tag{15}$$

The energy release predicted by this expression can be compared with that measured using an approximate energy calibration. The apparent "energy," as determined with acoustic emission equipment, from the bursting of bubbles formed by the reaction of a drop of dilute hydrofluoric acid with a beryllium sample was measured.[57] The actual energy released by the bubble burst was calculated and compared with the "energy" as measured by the acoustic emission equipment, thereby giving an approximate calibration for the "energy" as measured by the acoustic emission equipment. From this calibration, a typical acoustic emission burst in beryllium results from an event releasing approximately 5×10^{-3} ergs at a stress level of 200 MN/m^2. The dislocation density was on the order of $5 \times 10^8/cm^2$ and $b = 2.3 \times 10^{-8}$ cm. If all the dislocations within a cube of side d are presumed to move nearly simultaneously, then a cube 7 μm on a side is required for an average burst. Such a cube would contain 0.2 cm of dislocation line at the measured dislocation density. The average grain size in the beryllium tested was 19 μm, thus the calculated volume of yielding material is physically plausible.

For the first time, an absolute measurement of the surface displacement caused by dislocation motion has been reported.[14] The acoustic emissions involved were the highest amplitude bursts generated during deformation of a high-purity Al-5.55 percent Zn-2.5 percent Mg alloy aged 100 hours at 120°C. The time history of the source area was calculated from the measured surface displacement versus time using the model leading to Eq. (7). Assuming multiple loops growing cooperatively to the boundaries of the average size grain (420 μm), more than 10,000 dislocations were calculated to have moved cooperatively to produce the largest acoustic emission bursts. (At maximum radius, the length of dislocation loop would be more than 1,300 cm).

Several other calculations have been made which may help define the mechanism for energy release by dislocation motion. Natsik and Chishko[94] calculated the energy E_{an} produced by the annihilation of two dislocations. For a pair of screw dislocations

$$E_{an} = \frac{\rho\, b^2\, v^2}{8\pi} \ln \frac{\ell}{b} \tag{16}$$

where ρ is the density, b is the Burger's vector, v the relative velocity at annihilation (they suggest this is between 0.1 and 1.0 times the shear wave velocity), and ℓ is the size of the crystal. The magnitude of the energy released by annihilation can be compared to the energy release from yielding calculated above. For beryllium, taking $\ell = 7 \times 10^{-4}$ cm (the cube side determined above), $b = 2.3 \times 10^{-8}$ cm, $\rho = 1.8$ g/cm^3, and $v = 4.5 \times 10^5$ cm/sec (half the shear wave velocity), E_{an} is approximately 8×10^{-5} ergs per centimeter of annihilating dislocation line. Thus, about 120 cm of dislocation line must be annihilated to produce a 5×10^{-3} erg energy release. Therefore, the energy release from annihilation is two to three orders of magnitude less than that released from the yielding resulting from dislocation motion.

Accelerating dislocations emit energy in a way analogous to energy radiated by an accelerating charged particle. Kiesewetter and Schiller[95] proposed that the generation, motion, and arrest of dislocations from a Frank-Read source could be treated as a periodic process. They applied the expression derived by Eshelby for the energy radiated per second from a harmonically oscillating screw dislocation directly to the above process and obtained the result that the energy release per second E should be

$$E \propto v^2 L \ell, \tag{17}$$

where v is the average dislocation velocity, L is the total length of mobile dislocations, and ℓ is the average length of moving dislocation segments. Since $V_{rms} \propto \sqrt{E}$, this expression predicts the same dependence on dislocation velocity, segment length, and total length of moving dislocation as the expression of Hsu and Ono[74] (Eq. (14)). Ookawa and Yazu[96] have also calculated the radiation expected from a dislocation moving on the average at a constant velocity, but accelerated and decelerated by passing through local strain fields.

Cortellazze *et al.*[97] have calculated the expected value of the atomic displacement in a crystal from a uniformly moving edge dislocation. Natsik and Burkanov[98] calculated some characteristics of the elastic wave generated by an edge dislocation emerging at the surface of a crystal. Natsik and Chishko[99] have calculated the time dependence of the plastic deformation expected from the activation of a Frank-Read source, its free operation, and subsequent shutdown as the emitted dislocations pile up against some obstacle.

4. EFFECTS OF EXPERIMENTAL AND MATERIAL VARIABLES

The acoustic emission from dislocation sources is dependent on a number of experimental and material parameters not discussed in detail in the preceding sections. These parameters include strain rate, test temperature, purity, irradiation, crystal structure, stacking fault energy, prior thermomechanical treatment, grain size, sample size, and stress state. The effects of these variables will be discussed in the following sections.

4.1 Strain Rate

If the mechanism responsible for producing acoustic emission is not altered by a change in strain rate, then the only effect of a change in strain rate should be a corresponding change in the rate at which acoustic emission events occur. Thus, the burst or event rate should be proportional to strain rate, as observed experimentally.[51,56] Again, if the events themselves are not affected by strain rate, the power in the acoustic emission signal should be proportional to strain rate, which has also been observed experimentally.[6,53,72,100] Since the energy is proportional to $V^2_{rms}\,\Delta t$ (V_{rms} is the root mean square voltage of the acoustic emission signal over a time interval Δt), then V_{rms} should be proportional to $\dot{\epsilon}^{1/2}$, where $\dot{\epsilon}$ is the strain rate.[72,100] This proportionality has been observed in a host of different materials, both single crystals and polycrystals, at a number of purity levels.[23,95,100–106]

There are a few reports of other strain rate dependencies. Kishi and Kuribayashi[59] report $V_{rms} \propto \dot{\epsilon}$ for α iron and steel, and $V_{rms} \propto \dot{\epsilon}^{1.5}$ for zinc and cadmium polycrystals. Twinning is an important acoustic emission source in both zinc and cadmium. In suitably oriented, textured magnesium plate, two acoustic emission peaks which occur at different strains are observed.[107] The low strain peak (in the macroscopically elastic region) has been attributed to twinning and the peak at yield to dislocation motion. The height of the dislocation peak obeys the expected proportionality of V_{rms} to $\dot{\epsilon}^{1/2}$, but the height of the twinning peak has no consistent strain rate dependence.

For extruded Ti rod of a suitable orientation, the stress required to initiate twinning *decreases* with decreasing temperature from room temperature to −200°C.[107] In this temperature range, the acoustic emission shows a corresponding *decrease* with increasing strain rate.

The count rate is not simply related to the energy release rate. The dependence of count rate on plastic strain rate can be seen by substituting $C\,\dot{\epsilon}^{1/2}$ for $Vrms_m$ in Eq. (3). For example, the maximum count rate observed from 7075 aluminum increased by 1,000 times when the strain rate was increased by 100 times.

There is a higher level of agreement on the strain rate dependence of acoustic emission than on virtually any other experimental parameter. However, several precautionary notes are in order. First, for signals near the system noise level, the proper noise subtraction (see Eq. (4)) must be performed or an incorrect strain-rate dependence can be obtained.[23] Second, the acoustic emission source mechanism may be affected by strain rate which will lead to a different strain rate dependence. For example, the yield point load drop and its accompanying high level of acoustic emission in 5083 aluminum disappeared as the strain rate was reduced.[68] Similarly, Portevin-LeChatelier yielding and the acoustic emission which typically accompanies it can be suppressed either by slowing down or increasing the strain rate. As another example, the deformation mechanism in zinc shifts from slip to twinning as the strain rate is increased[50] which will also change the acoustic emission strain rate dependence. It also appears, from the evidence presented above, that twinning can be strain rate sensitive and produce peculiar acoustic emission strain rate dependencies.

4.2 Test Temperature

Test temperature is somewhat analogous to strain rate, in that changes in deformation mechanisms produced by altering the strain rate can also be produced by changing the test temperature. For example, changing the temperature changes solute atom mobility and can cause the appearance or suppression of serrated yielding or yield points. There have been several examples of this behavior observed during acoustic emission testing.[73–76] Hartman[71] has also observed differences between room temperature and 325°C in the acoustic emission from α-brass deformed with deadweight loading. Portevin-LeChatelier yielding was observed at both temperatures. At room temperature, most of the acoustic emission energy was released during the yield increments, while at elevated temperature, the acoustic emission occurred less regularly. Therefore, as with strain rate effects, care must be taken to separate activation of different acoustic emission mechanisms at different temperatures from the temperature dependence of a specific mechanism.

These considerations aside, there are large changes in the acoustic emission (attributed to dislocation motion) with temperature for some materials. Hsu and Ono[74] measured the rms acoustic emission from polycrystalline 99.99 percent Al, 99.99 percent copper, and 99.4 percent iron, Figure 9. For aluminum, the acoustic emission increased with increasing test temperature from –60°C to 25°C, and decreased at higher test temperature up to the maximum temperature tested, 300°C. Copper behaved similarly, with the maximum emission at 300°C (test temperature range –60°C to 600°C).

Similar behavior was also observed with iron, although the changes with temperature were smaller than for aluminum and copper. The maximum acoustic emission occurred at a test

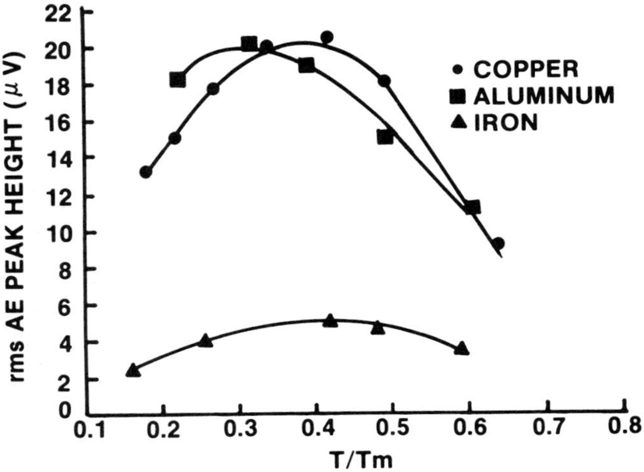

Figure 9. Height of rms acoustic emission peak near yield as a function of measuring temperature, divided by the melting temperature, for Cu, Al, and Fe. From Hsu and Ono, Reference 74.

temperature of 500°C (test temperature range 27°C to 800°C). Portevin-LeChatelier yielding was observed at 400°C. Correction of the measured values for increase attentuation at high temperatures did not alter the qualitative nature of the results. Kishi and Kuribayashi,[59] in contrast, reported an approximate linear decrease in peak rms near yield for copper from room temperature to 400°C, with similar results for aluminum and other fcc pure metals.

Hsu and Ono[74] propose that the temperature dependence arises from two competing factors. The viscous drag on moving dislocations increases with increasing temperature, leading to a decrease in dislocation velocity and lower acoustic emission (Eq. (7)). As the temperature is raised, Hsu and Ono propose that the internal stress distribution also changes to a larger period, which increases the dislocation glide distance and thereby increases the acoustic emission. The evidence for this proposed change is the increase in dislocation cell size with deformation at higher temperatures, although cell formation is not observed until strains well beyond the maximum in acoustic emission. Another possibility is that additional sources are available at higher temperatures from thermal activation. In addition, the C_1^2/C_2^3 term in Eq. (7) will increase with increasing temperature because the velocities decrease with increasing temperature. This is, however, a fairly small effect. For 1100 aluminum, the increase from this term is about 5 percent per 100°C near room temperature, using the data from Reference 108. (These data are for temperatures below room temperature, but the elastic modulii decrease linearly with temperature above 150°K.) Finally, temperature also affects the yield stress and the shape of the load-time curve, which both influence the acoustic emission (see also the effects of grain size, Section 4.9).

Hsu and Ono[75] have also measured acoustic emission as a function of temperature for a number of alloys. For Cu-10% Ni, Cu-30% Ni, Cu-10% Zn, and Cu-15% Zn alloys, the temperature dependence of the rms peak at yield was similar to pure copper. The maxima were at the following temperatures: Cu-10% Ni, 375°C; Cu-30% Ni, 500°C; Cu-10% Zn, ≈325°C; Cu-15% Zn, ≈400°C. The height of the peak increased with increasing alloy content. Portevin-LeChatelier yielding was observed in these alloys at temperatures near the temperature of maximum acoustic emission, but serrations were not observed until strains well beyond the location of the acoustic emission peak near yield. In a second group of alloys, Cu-30% Zn, Cu-40% Zn, Cu-7.5% Al, Ni-22% Fe, the acoustic emission versus test temperature also exhibited a maximum. However, the increase in AE with temperature and the maximum attained were higher, the maximum occurred at higher homologous temperatures, and acoustic emission dropped more sharply at higher test temperatures than in the first group of alloys. The maxima were at the following temperatures: Cu-30% Zn, 500°C; Cu-40% Zn, 500°C, Cu-7.5% Al, ≈450°C; Ni-22% Fe,

800°C. Substantial Portevin-LeChatelier yielding was observed, but at test temperatures below the test temperature at which the maximum acoustic emission at yield occurred. A strong second peak was also observed, centered at strains between 5 and 15 percent, for a limited range of test temperatures which were also below the test temperature for the maximum in the first peak.

Hsu and Ono attribute the acoustic emission behavior in the first group of alloys to the same mechanisms as in pure metals, with an additional contribution due to breakaway from solute atmospheres. They attribute the higher emission and stronger temperature dependence in the second group of alloys to breakaway of dislocations from pinning by short range order with an additional significant contribution from breakaway from solute atmospheres.

Hsu *et al.*[76] have also measured the acoustic emission from 304L stainless steel between room temperature and 675°C. The peak rms acoustic emission level near yield increased substantially throughout the range of temperatures measured with the peak height at 675°C being about 20 times higher than at room temperature. Serrated yielding was observed above 400°C. The authors postulated that a change in slip mode from planar to non-planar slip with increasing temperature (because of increased stacking fault energy) was responsible for the increased emission near yield. The argument is similar to that advanced above for pure metals; non-planar glide leads to the formation of dislocation cell structures enclosing relatively dislocation-free zones. The dislocation packet glide distance in these dislocation-free zones is larger than in structures where cells are not formed during deformation, which leads to increased acoustic emission. See Section 4.6 for a further discussion of this model. Kishi and Kuribayashi[59] also report a steady increase in the height of the rms peak at yield in 304 stainless steel from room temperature to above 500°C with similar behavior in α-brass.

Kishi and Kuribayashi[59] reported a strong maximum in the

height of the acoustic emission peak near yield for mild steel at about 300°C. Okajima and Ono[41] also report a substantial increase in the portion of the acoustic emission attributed to dislocation motion in A533B steel from –150°C to 150°C, which is consistent with the above result for mild steel. Finally, Kishi and Kuribayashi[59] report a precipitous drop in the rms acoustic emission peak near yield in 5056 aluminum (an Al-5.2 percent Mg alloy) with increasing temperature from room temperature to 500°C.

4.3 Sample Purity

Providing that solute locking of dislocations which leads to a yield point and/or serrated yielding does not occur, it appears that the addition of solute reduces the acoustic emission observed near the onset of plastic flow. Kuribayashi and Kishi[64] observed a reduction in the height of the rms peak near yield in Al-Mg single crystals such that the peak height was proportional to $C^{-1/2}$ were C was the magnesium concentration (0 – 3 atomic percent). They reported similar behavior for Cu-Al and Al-Si systems. Scruby *et al.*[53] also observed substantially less acoustic emission from large-grained (2900 μm) Al-1.3% Mg samples compared to 99.999% Al of comparable grain size. Kiesewetter and Schiller[95] report a substantially lower acoustic emission peak at yield for 99.5% Al polycrystals compared to 99.99% Al polycrystals, however, the grain sizes were also different (200 μm for the pure aluminum versus 50 μm for the less pure material). The authors have also observed that the acoustic emission from 99.9995% Al was greater than from 1100 aluminum; however, the grain size in the two materials was also different. Hatano[109] reported a drastic decrease in acoustic emission from Cu-Al single crystals with increasing aluminum content (0-1.15 weight percent). For the Al-Zn system, Kuribayashi and Kishi[64] report that the $C^{-1/2}$ dependence of acoustic emission on solute concentration was not observed, in contrast to the Al-Mg, Cu-Al, and Al-Si systems.

The difference was proposed to arise from a lower attraction of Zn atoms to dislocations compared to the other solute atoms.

If the material or experimental conditions are such that solute locking of the dislocations occurs, then the addition of solute can cause very large increases in acoustic emission compared to that from the pure base metal. Scruby *et al.*[53] observed a large increase in acoustic emission from Al-1.3% Mg samples with a grain size near 80 μm compared to 99.999% Al of similar grain size. A yield plateau and Portevin-LeChatelier yielding were observed for the Al-Mg alloy samples in this grain size range. Acoustic emission levels much higher than in pure aluminum have also been reported[68] in 5083 aluminum (4.5 percent Mg) under conditions where a yield point and serrated yielding are observed. Similar results have been reported[69] for 5086 Al (3.5 percent Mg).

Because of the extreme sensitivity of the deformation properties of bcc metals to small concentrations of interstitial impurities, one would also expect acoustic emission to be sensitive to interstitial impurities. Higgins and Carpenter have shown that, under conditions appropriate for pinning of the dislocation line by interstitial impurity atoms, the acoustic emission activity increased with increasing interstitial concentration.[37]

4.4 Irradiation

Irradiation produces defects which may act as breakable pins and increase the probability for dislocation breakaway avalanches (increasing the acoustic emission), but the defects may also serve to reduce the glide distance of the avalanches and/or their velocity (reducing the acoustic emission). Which effects predominate will probably depend on the material and possibly the dose. Experimentally, there is a divergent response in the effect of prior irradiation on the acoustic emission from different materials.

The burst rate from LiF was found to increase somewhat after x-ray irradiation.[58] The amplitude of the bursts was found to be about 10 times larger after irradiation. Internal friction measurements proved conclusively that the dislocations were effectively pinned by the defects produced by x-ray irradiation.

Tests in A302 Grade B reactor vessel steel exposed to 2×10^{18} neutron/cm^2 (energy > 1 MeV) at 290°C showed an increase in the total counts through yield of about four times, compared to unirradiated material.[110] A tenfold increase in total counts through yield was reported in the same material after irradiating to a dose of 1.2×10^{19} neutrons/cm^2 at the same temperature and neutron energy.[111] An increase in counts during the burst testing of slotted zircalloy tubes after irradiation has also been reported.[112]

Nakasa[84,85] has investigated the effects of 2.7×10^{17} neutrons/cm^2 (energy > 180 keV) on 99.999 percent Cu, 99.99 percent aluminum, 99.99 percent iron, 99.8 percent vanadium, vanadium-600 ppm boron, and vanadium-700 ppm carbon. A large reduction in the count rate at yield (250 times) and a reduced average signal amplitude was reported in copper. Smaller reductions in count rate at yield were observed in vanadium-boron alloy (3 times), vanadium (2.4 times), and aluminum (1.3 times). The count rate appeared to increase slightly for iron and vanadium-carbon alloy.

4.5 Crystal Structure

The effects of crystal structure on the acoustic emission from dislocation sources are difficult to evaluate due to the interplay of other material parameters. Care must be taken to try to sort out effects of grain size, stacking fault energy, impurity effects, etc., in order to single out effects which are primarily due to crystal structure. The difficulties in doing this are clear, and for this reason little definite data exist on the effects of crystal structure. An additional major difficulty is that twinning is much more common in some crystal systems than in others. Imaeda *et al.*[103] have investigated the acoustic emission from a number of metals of various crystal structures when deformed in tension. The crystal structures of metals tested included fcc, bcc, bct, and hcp. The metals were all polycrystalline of commercial purity. Some general observations from this work and other acoustic emission data are possible for pure metals of different crystal structures. Metals having an fcc structure are characterized by a single peak in the acoustic emission near yield of the material. In many materials the peak is a smooth curve in either count rate or V_{rms}. The peak is usually quite narrow and falls off rather rapidly with increasing strain, with essentially no acoustic emission at large strains. Clearly, this behavior can be altered in fcc alloys; refer to the previous comments concerning 5083 aluminum which is a fcc alloy. Metals having a bcc structure are also characterized by a peak in the emission at yield; however, many bcc metals also show significant acoustic emission during the yield point elongation (the flat portion of a stress-strain curve following a yield point). After the yield point elongation, the acoustic emission decreases and, like fcc materials, little emission is observed at large strains. Although the peak amplitudes of acoustic emission signals from fcc and bcc metals are essentially the same, the acoustic emission measured either by counts or V_{rms} is much more sporadic or nonuniform from bcc metals than from fcc metals. Acoustic emission curves for bcc metals are typically very broad, with peaks and scatter. One would expect that purity levels would be of greater importance to bcc metals when making comparisons of data due to the known importance of interstitial impurity locking of dislocations.

The acoustic emission from hcp metals is, in most cases, significantly greater than that measured in either fcc or bcc metals. The peak amplitudes of the emissions from hcp metals are also found to be larger. Twinning is common during deformation of hcp metals and the emission observed may consist largely of emission from twin formation. An extremely wide variation of acoustic emission behavior for hcp materials has been reported. For example, in zinc the acoustic emission has been found to start at yield and increase with deformation up to 10% at which the test was stopped. Similar data exist for titanium.[113] Testing of zirconium shows a maximum of the acoustic emission in the elastic region, a slight decrease, and then another maximum at approximately 3 percent strain, after which the acoustic emission rapidly falls to zero.[103] Testing of pure cast magnesium produces completely different acoustic emission results, depending on whether the sample is taken parallel or perpendicular to the plate.[114] In both cases, the emissions are large and highly sporadic, giving a wide band of scatter in the acoustic emission when measured either by counts or V_{rms}. Acoustic emission from dislocation sources generally reaches a maximum near yield, but an acoustic emission peak at higher strains has been attributed to dislocations in beryllium.[51] The burst rate peak is centered at about 1 percent plastic strain, and was shown not to result from precipitate frature, grain cleavage, twinning, or cracking, and not to be affected by impurity content. The peak was postulated to result from breakaway from pins generated by the deformation process and may arise from some peculiarity in dislocation behavior in beryllium. Acoustic emission during deformation has also been reported for ionic crystals of the NaCl-type structure.[58,115] These materials appear to behave, in most respects, in a manner very similar to that observed for fcc metals.

4.6 Stacking Fault Energy

Most of the early studies regarding the effects of stacking fault energy were associated with investigations of the Bauschinger effect. The Bauschinger effect is the phenomenon that when a metal which is initially loaded in one direction (for example, tension) beyond yield, unloaded, and is subsequently reloaded in the opposite direction (i.e., compression), then the stress required for plastic flow is less than if the sample were reloaded in the original direction. The parameter usually used to quantify the magnitude of the Bauschinger effect is the Bauschinger strain, which is defined as the excess strain which occurs during unloading and reverse loading (up to some specified reverse loading stress) over that which would have occurred had the specimen behaved elastically. This effect is believed to originate because dislocation structures (the simplest being dislocation pileups) are formed during unidirectional deformation, which are easier to move in the reverse direction than to move farther in the original direction. The stacking fault energy is important in this phenomena because the ease of cross slip around obstacles is greater in high stacking fault energy materials. Because of this fact, the pileups in high stacking fault energy materials will be smaller and less frequent, leading to a smaller Bauschinger effect.

Early results correlated the magnitude of the emission during unloading to the magnitude of the Bauschinger effect, however, the major part of the acoustic emission observed during unloading was later discovered to have originated in the load train.[116] Subsequent work indicated, however, that some acoustic emission did occur during unloading. More recently, Williams and Emerson[117] measured acoustic emission during loading and unloading in the macroscopically elastic region of polycrystalline magnesium AZ31B-F alloy and correlated the amount of unload emission with the Bauschinger strain in the unloading cycle.

Averbukh and Vainberg,[118] and Kishi *et al.*[119,120] have reported acoustic emission during loading, unloading, and reverse loading portions of Bauschinger effect tests. They observed that acoustic emission during the unloading-reverse loading portion of the test began to increase when plastic flow began in the reverse direction, reached a maximum, and then declined. The peak remains, although it decreases in size, during repeated load-unload-reverse load cycles. Kishi *et al.*[119]

measured this peak in 99.5 percent aluminum, OFHC copper, α-brass, type 304 stainless steel, and aluminum-1.2 percent silicon. They related the peak to the back stress generated in the material by dislocation structures created in forward deformation. The magnitude of the back stress calculated from the acoustic emission peak during reverse loading, the changes in the stress at which the reverse loading peak occurred versus the peak forward stress, and the changes of the back stress calculated from the acoustic emission data with aging of the aluminum-silicon alloy were all consistent with known behavior of the Bauschinger strain with increasing stacking fault energy.

More recently, consideration has been given to the effect of stacking fault energy on acoustic emission during initial loading because stacking fault energy significantly influences the character of dislocation motion. Hatano[109] has measured the acoustic emission from copper, copper-aluminum alloys, α-brass, and 304 stainless steel, and suggested that the acoustic emission decreases with decreasing stacking fault energy. The bulk of the change was between copper (taken to be 41 ergs/cm^2 stacking fault energy) and copper-1.5 percent aluminum (33 ergs/cm^2), while aluminum (~200 ergs/cm^2) was essentially the same as copper. Hsu, Ono, and Hatano[76] postulated a mechanism by which changes in stacking fault energy could alter the amount of acoustic emission observed. As the stacking fault energy is increased, cross slip becomes easier. In materials with easy cross slip, the dislocation structure formed after significant deformation consists of dislocation cells and tangles with relatively dislocation-free areas outside the tangles and cell walls. Dislocation packets traveling across these open areas would have a longer glide distance than if the dislocations were more uniformly distributed (as in low stacking fault energy materials). The longer dislocation glide distance produces increased acoustic emission. This concept was used to explain the temperature dependence of the acoustic emission in 304L stainless steel, as discussed in Section 4.2.

One difficulty with this concept is that the dislocation cell structure does not develop until extensive plastic flow has occurred, usually more than about 5 percent. The peak in acoustic emission occurs at the onset of plastic flow, often at plastic strains less than 0.2 percent, and it is unclear that subsequent cell formation is important at low strains. In addition, as noted by Hsu *et al.*[76] several metals and alloys do not fit into the pattern of decreasing acoustic emission with decreasing tacking fault energy. Imaeda *et al.*[103] found no consistent relation between the rms voltage of the acoustic emission peak at yield and the stacking fault energy for the fcc metals silver, copper, aluminum, and nickel.

Heiple and Carpenter[86] have proposed that stacking fault energy is one of the major parameters that influence the effect of aging precipitates on the acoustic emission during deformation. Cuttable precipitates are proposed to be more effective sources of dislocation avalanches in difficult cross slip systems (low stacking fault energy) because dislocations tend to pile up behind the precipitates, rather than cross slip around them, as in easy cross slip systems. Data from precipitation hardening systems are discussed in Section 4.8 (Heat Treatment).

4.7 Prior Mechanical Work

As indicated in Section 3.5, several explanations have been offered for the reduction in acoustic emission from dislocation motion observed at strains beyond yield in metals. The proposed mechanisms include a reduction in the glide distance of the dislocation packets by the obstacles (mainly other dislocations) produced during the preceding deformation, a shift in the energy spectral density of the acoustic emission because of a reduced glide distance, and a reduction in the rate of formation of dislocation avalanches because some of the potential sources were activated earlier in the deformation process. All these mechanisms are consistent with the common observation that the acoustic emission behavior of a material can be substantially influenced by its prior working history, provided that it is not recrystallized after working. Even if it is recrystallized, a secondary effect can be observed through the dependence of acoustic emission on grain size because the grain size depends, in part, on prior working history. The most notorious example of a difference produced by prior work is the acoustic emission behavior of 7075-T6 and 7075-T651 aluminum. Aluminum in the T651 condition has been subjected to a stress relief stretching operation involving up to 3 percent plastic strain following the solution heat treatment in order to improve flatness. This mechanical working suppresses almost completely the small acoustic emission peak observed at yield from dislocation processes in the T-6 temper.[47] Another example is provided by JBK-75, a precipitation hardenable austenitic stainless steel similar to A286. This material produces substantial acoustic emission in the recrystallized and aged condition,[1] but essentially no emission in the forged and aged condition. The low acoustic emission in the forged and aged condition is presumably because of the high dislocation density and consequent short glide distance. It is therefore clear that the mechanical history of a material can play a major role in its acoustic emission behavior. Mechanical working can also have a substantial effect on non-dislocation sources of acoustic emission. It affects the size, shape, and distribution of inclusions and second phases which are known to produce acoustic emission in a number of systems. For example, the acoustic emission from cold-rolled mild steel is significantly different from hot-rolled mild steel because of the breakup of the cementite in the pearlite grains during cold-rolling.[121]

4.8 Heat Treatment

Heat treatment affects the grain size, the dislocation density and distribution, the solute atom distribution, and the precipitate type, size, and distribution (in age-hardening systems).

These effects have been shown or postulated to affect acoustic emission from dislocation motion. There are, of course, many additional effects of heat treatment which include, among others, changes in the type, distribution, and geometry of intentional second phases (such as cementite in iron), changes in crystal structure, or changes in inclusion distribution (such as the concentration and size of carbides in grain boundaries). These latter effects, while they may alter the acoustic emission from dislocation motion, will not be discussed in detail further.

Cooling rates from the annealing temperature have been shown to have a substantial effect on the observed acoustic emission, even in metals which are not age-hardenable. Frederick[87] has reported a large reduction in the total counts measured from 99.99 percent polycrystalline aluminum which had been quenched into ice brine from 630°C, compared to the same material which had been air-cooled. The difference was attributed to quenched-in vacancies. Borchers and Tensi[122] reported a 15-times reduction in the count rate of the acoustic emission peak at yield in AlMg3 (former German standard designation, the actual composition was 2.3 percent magnesium, 0.49 percent iron, 0.1 percent silicon, which is similar to 5052 aluminum) for samples water-quenched from 400°C, compared to samples furnace-cooled over 24 hours. A smaller reduction (approximately 4 times) was observed from 99.8 percent polycrystalline aluminum which had been water-quenched from 580°C, compared to material furnace cooled from the same temperature over 24 hours. They attributed the observed differences to different solute atom concentrations along and around the dislocations. This explanation is consistent with the relative difference between the fairly pure aluminum and the aluminum-magnesium alloy. Heiple[68] has reported a reduction of 60 times in V_{rms} of the acoustic emission at yield from 5083 aluminum in the solutionized and quenched condition, compared to the solutionized and slow-cooled condition.

In beryllium, rapid cooling is believed to create substantial numbers of mobile dislocations because of the large thermal stresses generated by the elastic anisotropy of beryllium.[80,123] The mobile dislocations are believed to respond easily to applied stress and be unlikely to form avalanches. This explains the observed reduction of the burst rate at yield by a factor of 2 to 4 in powder source beryllium air-cooled from 800°C, compared to the same material furnace cooled at 75°C/hour.[51] Similar effects may occur in other, more isotropic, materials.

Precipitates formed in age-hardening alloys clearly affect dislocation motion during deformation; that is the reason strengthening occurs. Thus, precipitates would also be expected to alter the acoustic emission, and such an effect has been observed for many precipitation strengthened alloys. The effect of precipitates has been viewed from the standpoint of their effect on the glide distance of the dislocation packets and their potential as breakable pins which lead to the formation of dislocation avalanches.

Frederick[87] observed a reduction in acoustic emission after aging in 2024 aluminum (with a modest increase in the overaged condition), and this result was interpreted by Agarwal, Frederick, and Felbeck[16] on the basis that the precipitates decreased the glide distance of the dislocations and the length of the source. Schmitt-Thomas *et al.* have studied in detail the acoustic emission as a function of aging in the German AlCuMg2 alloy[124] (werkstoff number 3.1355, an alloy essentially identical to 2024 aluminum) and in the alloy[125] AlCu4.3. They interpreted their results on the basis of a restriction in the glide distance of the dislocations by the precipitates formed during aging. This view does not account for an increase in acoustic emission when precipitates have been formed by aging, as has been observed in a number of alloys.

Scruby *et al.*[14] have observed a substantial increase in acoustic emission from an aluminum-4 percent copper alloy for aging times which precipitate fine, cuttable GP zones, Figure 10. Fenici *et al.*[106] have also reported a maximum in

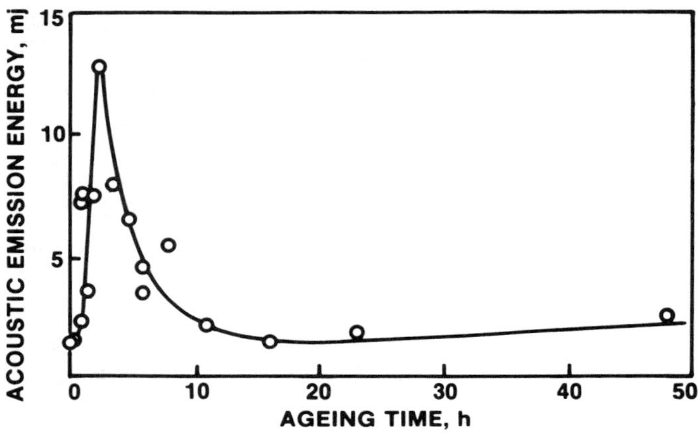

Figure 10. Total acoustic emission from the region near yield for Al-4% Cu as a function of aging time a 170°C. From Scruby *et al.*, Reference 14.

acoustic emission after short aging times (well before maximum strength is achieved) in an Al-4% Cu alloy. For aluminum-5.4 zinc-2.5 magnesium-1.6 copper alloys, a large increase in emission near yield was observed in the peak-aged condition. Scruby and coworkers propose that cuttable precipitates favor strain localization and the formation of coarse slip bands. The passage of the initial few dislocations, cutting the precipitates, forms a plane of weakness. The formation of a plane of weakness favors strain localization and the passage of many dislocations to form a slip band, which coarsens by cross slip. This process is proposed to increase acoustic emission because of the long glide distance and many dislocations involved. Transgranular slip bands were, in fact, observed in the peak aged condition in the aluminum-5.4 zinc-2.5 magnesium alloy. When the precipitates grow and become stronger, then dislocations tend to pass them by cross slip or Orowan bowing, and the strain localization does not occur.

Heiple, Carpenter, and Carr[86] instead proposed that the major effect of precipitates was to form (under favorable circumstances) initiation sites for dislocation avalanches. In alloy systems where cross slip is difficult (low stacking fault energy, high elastic modulus), the precipitates can serve as breakable or cuttable pins if they are not too strong, which is a favorable condition for the formation of dislocation avalanches, and leads to higher acoustic emission near the onset of plastic flow. With longer aging times, the precipitates grow larger and stronger, and no longer serve as breakable pins. The larger precipitates also more effectively restrict the glide distance of any avalanches which form. Precipitation strengthening stainless steels fall into this class of alloys, and an increase in the height of the rms peak near yield was observed after short aging times in JBK-75 (an alloy similar to A286), an experimental austenitic stainless steel (KHB) hardened with a beryllium-containing precipitate, and Incoloy 903. The height of the acoustic emission peak at yield for JBK-75 and KHB as a function of aging time are shown in Figures 11 and 12. No acoustic emission peak was seen in 17-10 P, where the hardening precipitate was an uncuttable carbide.

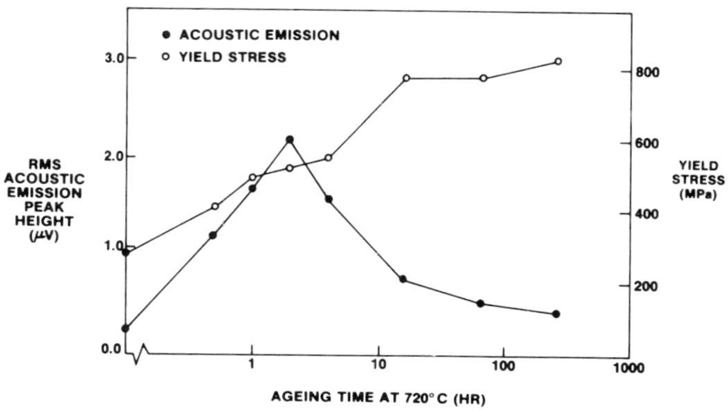

Figure 11. Yield stress (0.2%) and magnitude (referred to transducer output and corrected for noise) of rms acoustic emission peak near yield in JBK-75 stainless steel as a function of aging time. From Heiple, Carpenter, and Carr, Reference 86.

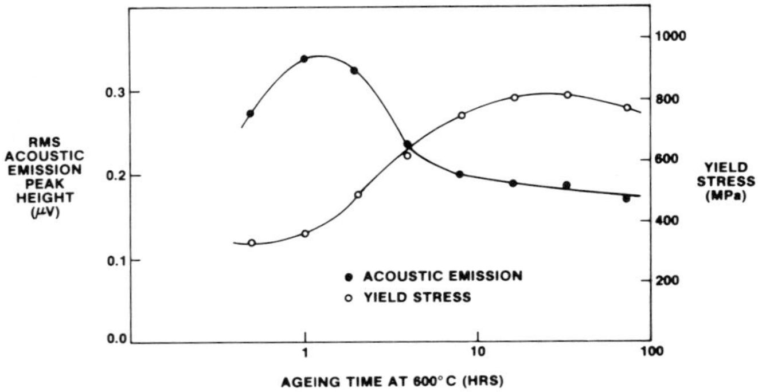

Figure 12. Yield stress (0.2%) and magnitude (referred to transducer output and corrected for noise) of the rms acoustic emission peak near yield in a beryllium-strengthened stainless steel as a function of aging time. From Heiple, Carpenter, and Carr, Reference 86.

In easy cross slip systems (high stacking fault energy, low elastic modulus), the precipitates do not as easily serve as breakable pins because dislocations tend to cross slip around them, rather than create pileups. Thus, aging would not usually increase acoustic emission in easy cross slip systems, although a narrow range of precipitate size and strength may exist which would create favorable conditions for breakaway avalanches. Precipitation hardening aluminum alloys fall into this class, and a decrease in acoustic emission compared to the solution treated and quenched condition was observed for all aging conditions tested in 7075, 6061, and 2219 aluminum and sterling silver. The height of the acoustic emission peak near yield as a function of aging in 7075 aluminum is shown in Figure 13. No aging treatment tested in these commercial alloys resulted in a large increase in acoustic emission like that observed by Scruby *et al.*[14] in aluminum-4% copper. Failure to observe such an increase may be a result of the limited number of aging treatments tested in the commercial alloys.

Precipitation also removes atoms from solid solution, and if the solute atoms had led to Cottrell atmosphere pinning

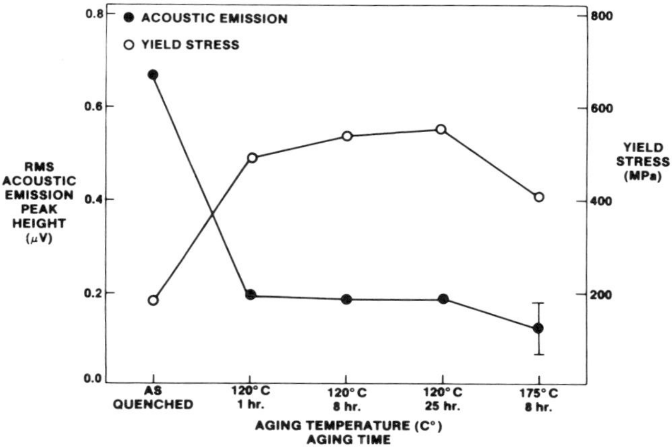

Figure 13. Yield stress (0.2%) and magnitude (referred to the transducer output and corrected for noise) of the rms acoustic emission peak near yield in 7075 aluminum as a function of heat treatment. The uncertainty in peak height in the overaged condition is due to overlap from a second acoustic peak centered at about 1.6% plastic strain. From Heiple, Carpenter, and Carr, Reference 86.

with consequent serrated yielding, then removing these atoms from solution by forming precipitates may significantly reduce the acoustic emission. This effect has been observed in 7075 (Figure 13) and 6061 aluminum alloys. In a detailed aging study, an increase (Figure 14) in acoustic emission at yield has been observed by Schmitt-Thomas *et al.*[125] in AlCu4.3 for aging treatments which produce reversion (re-solution during artificial aging of precipitates formed either during the quench from a solutionizing anneal, or during room temperature aging). They attributed the rise to increased glide distance of the dislocations, but an alternative explanation is that the reversion made more solute available for Cottrell locking.

Whether aging precipitates increase acoustic emission (under some conditions) because they form sites for dislocation

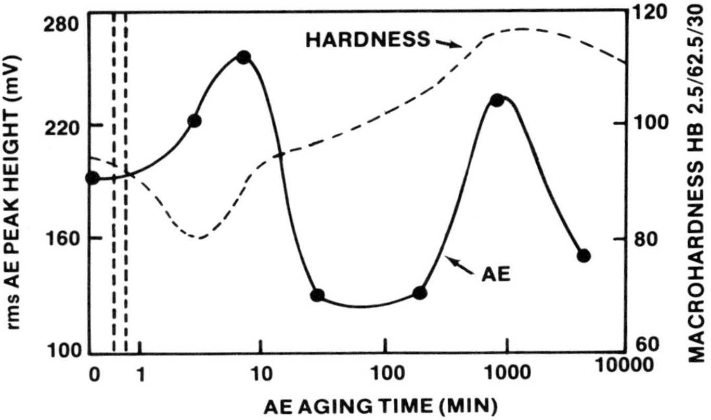

Figure 14. Hardness and height (corrected for noise) of the rms acoustic emission peak near yield in AlCu4.3 alloy as a function of aging time at 150°C. From Schmitt-Thomas, Tensi, and Zeitler, Reference 125.

breakaway avalanches, or because they promote flow localization into slip bands (or both) is not established. One useful experiment would be to determine if slip bands generally form at low plastic strains ($\lesssim 0.2$ percent, where the majority of the acoustic emission is observed) in alloys in which the acoustic emission is increased by prior aging but not in alloys in which the acoustic emission is decreased by prior aging.

The sequence of precipitates (solid solution $\rightarrow$ GPI $\rightarrow$ GP II $\rightarrow \gamma' \rightarrow \gamma$) formed during aging of copper-beryllium alloys is similar to the aging sequence in the precipitation hardening beryllium stainless steel KHB discussed above. The stacking fault energies in the two alloys are also comparable. Masuda[126] has measured the acoustic emission as a function of aging treatment for copper-1.65 percent beryllium and copper-1.85 percent beryllium alloys. The changes in the height of the rms acoustic emission peak near yield with aging are similar to those observed for KHB. The acoustic emission increased from that in the solution-treated condition after initial aging treatments, which formed GP(1) (and probably some GP(2))

Guinier-Preston zone precipitates. Further aging resulted in a lower peak near yield. The acoustic emission peak disappeared when sufficient aging was performed to produce maximum strength (γ' precipitates). The level of acoustic emission was low for all aging treatments.

Masuda has also reported[126,127] acoustic emission as a function of aging for three other copper-base precipitation hardening alloy systems, including copper-cobalt, copper-titanium, and copper-nickel-silicon (Corson) alloys. For copper-1.03 percent cobalt and copper-3.15 percent cobalt, there was a decrease in the rms peak height at yield from that in the solution-treated condition after an initial aging treatment, a rise in the height of the acoustic emission peak to a maximum near the maximum strength condition, and a decline with further aging. The peak was higher (about five times) in the solution-treated condition for the 1.03 percent cobalt alloy compared to the higher cobalt alloy. However, in the maximum strength condition, the peak near yield was twice as high in the 3.15% Cu alloy. The initial precipitates to form in this alloy are coherent, spherical, fcc particles rich in cobalt.[128] They grow with increased aging and become incoherent only after severe overaging. The initial precipitates are small (~30A for 3 minutes at 650°C for a Cu-3.1% Co alloy), and have only grown to 100-150A by the maximum strength (peak aged) condition.

For the copper-2.8 percent titanium alloy, the acoustic emission peak near yield increased in height with aging from the solution-treated condition, with the maximum height occurring in the slightly underaged condition. Increased aging produced a metastable β' phase which twinned extensively, resulting in intense acoustic emission beyond yield. The acoustic emission near yield in this alloy was stronger than in any of the other copper alloys tested. Precipitation in this alloy occurs by spinodal decomposition of the supersaturated solid solution.[129,130] Upon initial aging, there is a redistribution of the Ti along ⟨100⟩ directions in the copper lattice. This produces a periodic composition modulation, the amplitude of which increases with increased aging. The composition gradients also steepen at the interface of the Ti-rich rigions with increased aging. Further aging results in transformation of the Ti-rich regions to metastable β'. Maximum strength occurs after formation of the β'.

Three copper-nickel-silicon alloys were also tested; copper-7.99% nickel-1.48% silicon, copper-6.01% nickel-1.35% silicon, and copper-3.98% nickel-0.9% silicon. The acoustic emission peak height at yield increased with increased aging from the as-quenched condition and reached a maximum prior to the maximum strength aging treatment. The hardening precipitates in this alloy system are large Ni_5Si_2 intermetallic particles. These particles had an average diameter of 0.4 μm for the aging treatment producing maximum emission in the copper-3.98% nickel-0.9% silicon alloy, and were over 0.1 μm in diameter in the as-quenched condition.

Heat treatment effects have also been reported[131] in an aluminum-3 percent magnesium alloy, although this is not an age-hardening system. There was a large increase in the maximum rms near yield with longer aging treatments following a solution heat treatment. The level of acoustic emission was high, in agreement with results from the similar commercial alloys 5083[68] and 5086.[69] The height of the observed maximum in rms acoustic emission also depended on cooling rate after aging. The increased acoustic emission after aging was attributed to enhanced segregation of solute to dislocations and resultant increased solute locking.

Steels are commercially the most important heat-treatable alloys. The microstructural changes resulting from heat treatment of steels can be very complex, and sorting out which microstructural variables are responsible for changes in acoustic emission from dislocation motion is not an easy task. Most of the acoustic emission studies on the effects of heat treatment on steels have been aimed at effects on the emission during crack propagation. Although there is some contention that the source of acoustic emission (at least for some steels) during crack propagation is the plastic zone ahead of the crack (see, for example, Reference 132), acoustic emission from crack growth is outside the scope of this review.

Landy and Ono[133] have systematically investigated the acoustic emission from A533B steel with both bainitic and martensitic microstructures as a function of isochronal tempering treatments of 24 hours between 300°C and 750°C. They observed strong variations with tempering treatment in the height and shape of the rms acoustic emission peak near the onset of plastic flow, as well as the stress at which acoustic emission was first detected. The maximum in the rms acoustic emission peak near yield for the tempered martensite occurred for tempering temperatures between 200 and 300°C. They correlated their results to changes in initial dislocation density, pinning strength, and mean glide distance.

Rao and Gopal[134] have measured the acoustic emission from A533B steel for 1 hour tempers at temperatures between 400 and 675°C following austenitizing (900°C, 1/2 hour) and water quenching. The maximum count rate near yield occurred for tempering at 510°C and below, where ϵ-carbide plate-like precipitates were formed. Higher tempering temperatures produced lower count rates near yield even though a yield point and yield serrations were observed. Spheroidal carbides (cementite and alloy carbides) replaced the ϵ-carbide after the higher temperature tempers. Rao and Gopal attributed the acoustic emission after low temperature tempering to fracture of the ϵ-carbides, even though the acoustic emission maximum occurred at very small plastic strains. Sano and coworkers[105] have presented strong evidence that the acoustic emission from steels with a similar microstructure arises from dislocation motion.

Scruby *et al.*[11] have also measured the acoustic emission during deformation from steel of the A533B composition with a range of microstructures produced both by different cooling rates from the 1000°C austenitizing anneal and by isothermal tempering at 650°C of water-quenched samples. A small amount of emission near yield was observed in the samples air and furnace cooled from austenitizing. A larger acoustic emission peak occurred for tempering times of 8 hours or more. They proposed that acoustic emission near yield was detectable after slow-cooling and tempering treatments because these treatments increased the dislocation glide distance through a reduction in initial dislocation density, elimination of martensite lath boundaries, and an increase in carbide spacing.

The tempering temperature selected in this investigation appears to have been too high (even for short tempering times) to detect the large increase in acoustic emission after low temperature tempers observed by Landy and Ono [133] and Rao and Gopal [134] in A533B (see above), and also by Sano *et al.* [105] and Wadley *et al.* [135] (see below) in other tempered martensitic steels.

Sano *et al.* [105] have investigated acoustic emission during deformation as a function of 1-hour tempering treatments between 50°C and 750°C, following austenitizing and a water quench, Figure 15. The martensitic steels studied had compositions of iron-0.3 percent carbon, iron-0.4 percent carbon-1.5 percent silicon, and iron-0.3 percent carbon-1 percent vanadium. Both burst and continuous emission were observed. The large amplitude bursts, which occurred primarily in the as-quenched samples and samples tempered below 250°C, were associated with microcracking. The height of the rms continuous acoustic emission peak near yield increased strongly from the as-quenched condition with tempering temperature and was a maximum for a tempering temperature of 250°C for the plain carbon steel and about 300°C for the alloy steels. They associated changes in acoustic emission with carbide precipitation and growth. The maximum acoustic emission came after tempers which formed the most ϵ-carbide, as also reported by Rao and Gopal.[134] The ϵ-carbide is a coherent or semi-coherent carbide which transforms to cementite at higher

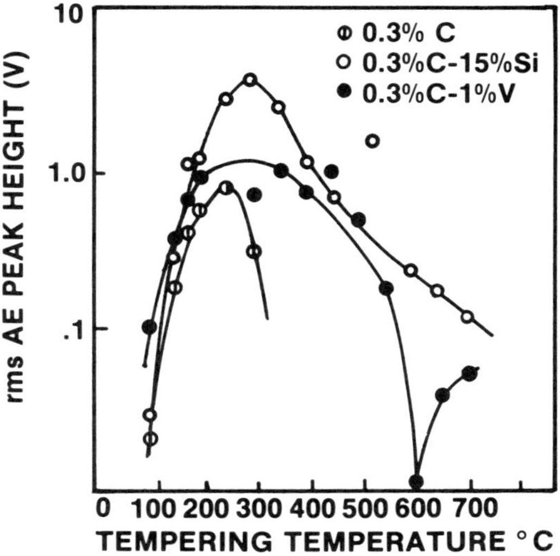

Figure 15. Height of the rms acoustic emission peak near yield for three quenched and tempered steels as a function of tempering temperature. (One hour tempers). From Sano *et al.*, reference 105.

tempering temperatures. Cracking of the ε-carbide was not detected by Sano and coworkers with transmission electron microscopy at low strains. Confirmation that dislocation motion was responsible for the observed acoustic emission came from internal friction measurements. A sharp rise in damping occurred after deformation to the stress at which the acoustic emission peak occurred. If the sample was then allowed to age at room temperature after deformation, and internal friction declined with time toward the pre-deformation value. They also observed a sharp minimum (Figure 15) in acoustic emission for the iron-0.3 percent carbon-1 percent vanadium steel for a 600°C tempering treatment which corresponds to secondary hardening by precipitation of a fine dispersion of vanadium carbides.

Wadley *et al.*[135] have also measured acoustic emission during deformation of a iron-3.5 nickel-0.2 carbon steel as a function of 100-minute tempering treatments between 50 and 650°C, following austenitizing at 1000°C and an oil quench. Both the height of the acoustic emission energy peak near yield and the total acoustic emission energy during the test were a maximum for tempering treatments between 200 and 300°C. This result is very similar to that described above by Sano *et al.* for iron-0.3 percent carbon steel. Wadley *et al.* propose that in the as-quenched condition and for tempering below 200°C, the high density of transformation-induced dislocations and small precipitates found in the martensite laths restrict the glide distance and velocity of dislocations so that little acoustic emission is detectable, in accordance with Eq. (7). Tempering between 200° and 400°C reduced the dislocation density, the density of original precipitates, and caused the formation of very small precipitates on martensite lath boundaries, prior austenite grain boundaries, and dislocations. Higher temperature tempering resulted in growth of these precipitates. They propose that the large precipitates formed at higher tempering temperatures restrict dislocation glide distance so that little acoustic emission is detectable. For the critical intermediate tempering temperatures (200-300°C), the fine precipitates may act as breakable (shearable) pins. The reduced mobile dislocation density (compared to lower temperature tempers) and lack of large, uncuttable precipitates (compared to higher temperature tempers) allow longer dislocation glide distances and produce substantial detectable acoustic emission.

4.9 Grain Size

Grain size would perhaps appear to be a simple material variable with a straightforward effect on acoustic emission from dislocation motion. It is not. Grain size changes produce the most complex melange of interacting effects on acoustic emission from dislocation sources of any variable considered in this paper. The experimental results available reflect the complexity of the problem: different experimenters using apparently similar materials have obtained diametrically opposite dependencies of acoustic emission on grain size.

A number of possible effects of grain size on acoustic emission are listed below. The relative importance of the different effects is not known.

1) For pure materials, the glide distance of dislocation avalanches near the onset of plastic flow will be proportional to the grain size. This effect tends to increase acoustic emission with increasing grain size (see Eq. (7)).
2) For pure materials, the most likely source of dislocations and dislocation avalanches is grain boundaries. Increasing the grain size reduces the grain boundary area and, hence, the number of possible avalanche sources. This effect tends to reduce acoustic emission with increasing grain size.
3) Reducing the grain size increases the yield stress. Therefore, dislocations in fine grained samples will be subjected to a higher stress when plastic flow begins than coarse-grained samples. The higher stress will increase dislocation velocity. For example, Gorman *et al.*[136] have shown substantial increases in dislocation velocity in aluminum with increased resolved shear stress. This effect tends to increase acoustic emission with decreasing grain size (see Eq. (7)). The reduction in flow stress may be great enough so that a high enough stress is never reached in coarse-grained sampled to activate sources which operate (at higher stresses) in fine-grained samples.[53]
4) Changing the grain size changes the shape of the load-time curve. Since virtually all tests have been constant cross head velocity tests, the strain rate near the onset of plastic flow where the maximum in dislocation-generated acoustic emission lies is not at all constant. The dependence of strain rate on time for this type of mechanical test depends on the shape of the load-time

curve, and thus on grain size. Since acoustic emission is strain rate dependent, the rate of acoustic emission will be affected.

5) For impure materials, the major source of dislocation avalanches may not be grain boundaries. The source could, for example, be breakaway from Cottrell atmospheres as when Portevin-LeChatelier yielding is observed. Thus, a grain size dependence different from pure materials might be expected, and has been observed.[53]
6) Reducing the grain size (at least for pure materials) decreases the glide distance and increases the dislocation velocity, as discussed above, which results in a shorter glide time for the dislocation packets. Fleischmann *et al.*[62] have shown a shorter glide time shifts the frequency distribution of the energy released by the moving dislocations to higher frequencies. Since measurements generally are made over a relatively narrow frequency band, this shift in the spectral distribution of the energy emitted will change the amount of acoustic emission detected.
7) Because of the varying annealing treatments required to obtain a range of grain sizes in a given material, the initial dislocation density will generally vary with grain size. The type and distribution of grain boundary dislocation sources, particularly ledges, may also vary with annealing treatment.

There have been many observations of the dependence of acoustic emission on grain size. The experimental observations are summarized in Table I. Some possible trends in the data may be noted. The maximum acoustic emission for the highest purity material appears to occur at the largest grain size or in single crystals. As the purity level decreases, or alloying elements (in solid solution) are added, the maximum acoustic emission apparently is observed at smaller grain sizes. The grain size at which the maximum in acoustic emission is observed may be less than the finest grain size achieved for a particular experiment. The grain size at which the maximum acoustic emission is observed also appears to depend on the frequency band in which the emission is measured, as predicted by Fleischmann *et al.*.[62]

Acoustic emission measurements during the deformation of single crystals have been made from the early days of acoustic emission testing. They were used to demonstrate that acoustic emission is not exclusively associated with grain boundaries, and that acoustic emission is generated in the bulk of the material and not just at the surface,[146] although surface oxides can substantially alter the observed acoustic emission (particularly in aluminum).[147]

Several characteristics of the acoustic emission versus strain are common to most of the measurements on single crystals. The acoustic emission begins at the onset of plastic flow, rises to a maximum, and then declines slowly with increasing strain. The peak (whether measured as count rate, rms, or energy rate) is much broader in strain than for polycrystalline material of comparable purity. The acoustic emission level falls slowly with increasing strain after the maximum. As indicated in Table I, some investigators observe the height of the peak to be greater in single crystals than polycrystals, while other investigators have obtained the opposite result.

There is, in addition, some disagreement as to the location of the peak in single crystal aluminum, which is the most commonly tested material. Kuribayashi *et al.*[64,148] observed the maximum at the transition between Stage I (easy glide) and Stage II (linear hardening) in 99.99 percent aluminum (20-200 kHz bandpass, 140 kHz resonant transducer). Scruby *et al.*[53] (99.999 percent aluminum, 100-1000 kHz bandpass, broadband transducer) report the maximum lies at about 4 percent strain in Stage II for their crystal, which did not exhibit easy glide. Shimizu *et al.*[149] (99.999 percent aluminum, 140 kHz resonant transducer) and Kiesewetter and Schiller[95] (99.99 percent aluminum, 140 kHz resonant transducer) found the maximum at the transition between Stage II and

TABLE I
Observations on the Dependence of Acoustic Emission on Grain Size

Material	Grain Size Range (μm)	Type of A.E. Measurement	A.E. Feature Measured	Frequency Bandpass (kHz)	Transducer Resonant Frequency (kHz)	Grain Size For Maximum A.E. (μm)	Reference	Notes
Group I: A.E. *increases* with increasing grain size								
99.99 Cu	15-250	counts	Total counts to 0.2%, 1% strain	100-300	150	largest	137	
99.99 Al	200-single xtal	rms	max. rms		140	single xtal	95	
>99.99 Al		rms	max. rms	100-300	170	largest	138	
99.999 Al	500-single xtal	A.E. energy	Total A.E. energy in test	100-1000	flat	single xtal	153	
Group II: A.E. has a maximum for grain size								
99.9 Al	22.72	some rms	max. rms		30	≈50	61	
99.99 Al	50-single xtal	rms	max. rms		900	2000	95	
99.99 Al	10-2000	counts	Total counts to selected stress	120-180		350-450	139	1,2
99.9 Cu	10-450	counts	Total counts to selected stress	120-180		<150	87	3
Fe-3% Si	35-1260	counts	Total counts to 90% lower yield stress		480	~400	92	4

Fe-3% Si	35-1260	counts	Total counts; 90% lower yield stress to 2.5% strain		480	smallest	92	4
1100 Al	40-120	counts	max. count rate		150-300	55-70	140	5
99.99 Cu	40-130	counts	max. count rate		150-300	70	141	
Al-1.3% Mg	25-2900	A.E. energy	Total A.E.	100-1000	flat	75	54	
Group III: A.E. *decreases* with increasing grain size								
"Pure Al"	21-11000	rms	max. rms	30-150	130-140	smallest	100,142,143	
99.9 Al	30-94	rms	max. rms		140	smallest	159	
99.9 Al	26-156	rms	max. rms		140	smallest	144	
99.9 Fe	900-single xtal	rms	max. rms	100-200	140	smallest	145	

Notes

1) The 10 and 20 μm grain sizes were subgrain sizes.
2) The total counts to several stresses were measured. Because the plastic strain required to achieve a given stress decreases with decreasing grain size (from the Hall-Petch relation), this type of measurement tends to underestimate the acoustic emission from fine-grained materials. The existence and grain size at which the maximum occurred were not, however, very sensitive to the stress level selected. The grain size at which the peak occurred shifted to a larger grain size when the count trigger level was raised.
3) Measurements made as in Note 2. The grain size at which the maximum occurred moved toward smaller grain sizes as the selected stress was increased.
4) The majority of the acoustic emission from silicon iron comes in the Luder's extension region (between the upper yield stress and 2.5 percent strain), so if the total counts (or rms peak height) were measured, the emission would decrease with increasing grain size in the grain size range tested.
5) The count rate maximum was at a grain size of 55 μm for a strain rate of 3.3×10^{-3}/sec and at 70 μm for 6.6×10^{-3}/sec.

Stage III (parabolic hardening). Toronchuk[150] (99.999 aluminum, 140 kHz resonant transducer) found the count rate maximum in Stage III for crystals oriented to favor easy glide.

The explanation for the varying location of the maximum may lie in measurements reported by Rouby and Fleischmann[63] in single crystal 99.99 aluminum. The found a broad acoustic emission peak whose location depended very strongly on the measurement frequency, Figure 16. For example, at a measurement frequency of 750 kHz, the maximum was at about 4 percent shear strain (near the end of Stage II), while at a measurement frequency of 2000 kHz the maximum was at about 18 percent shear strain (well into Stage III). They interpreted their results to indicate a rapid decrease in the duration of the dislocation glide event with increasing strain in Stage II, and a slow decrease in Stage III. The glide distance at the end of Stage II was calculated to be of order 12 μm.

Reported results for single crystal copper differ somewhat from those from aluminum. Hatano[109] and Mazzolai and Kim[151] reported a small acoustic emission peak at the onset

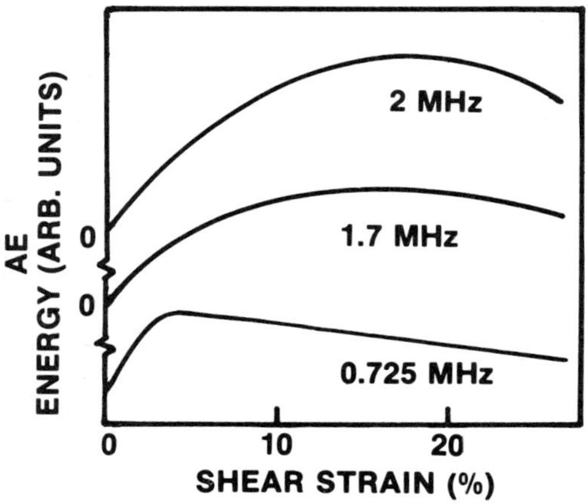

Figure 16. Acoustic emission energy detected at three different measurement frequencies as a function of shear strain for a 99.99% aluminum single crystal. From Rouby and Fleischmann, Reference 63.

of Stage I and a broad, larger maximum in Stage II. Kishi and Kurybayashi[59] and Jax and Eisenblatter[152] report a maximum in Stage I and a gradual decrease through Stages II and III.

Nakamura *et al.*[145] have reported the acoustic emission from 99.9 percent iron single crystals, bicrystals, and large grained polycrystals (100-200 kHz bandpass, 140 kHz resonant transducer). They reported a maximum at about 0.2 percent strain with a very gradual decrease with increasing strain. The peak for bicrystals was slightly sharper and higher. Acoustic emission began at lower strains in bicrystals than single crystals with a trend toward even lower strains in polycrystals of decreasing grain size. This trend was attributed to local plastic flow near grain boundaries before general yielding.

Kuribayashi *et al.*[148] reported differences in the height of the acoustic emission peak as a function of the initial orientation of single crystals of aluminum. They suggested that part of the difference was because of different strain rates on the primary slip plane as a function of orientation. For variously oriented single crystals of zinc, they correlated the rms acoustic emission peak height with shear strain rate on the basal slip plane. For copper, Hatano[109] reported the magnitude of the large peak in Stage II was nearly independent of initial orientation of the crystal.

4.10 Sample Size

If the events responsible for acoustic emission take place over a distance (or volume) small compared to the sample dimensions and the sources are homogeneously distributed in the sample, then the only effect of a change in sample volume should be a proportional change in the number of events activated at a particular strain (provided, of course, that other experimental variables such as strain rate are unchanged). These conditions should be satisfied for polycrystalline materials with a grain size small compared to the sample cross section, since the postulated mechanisms for acoustic emission are expected to operate within one or, at most, a few grains. Thus, burst rate or energy release rate should be proportional to sample volume, and V^2_{rms} should be proportional to sample volume. This will be true, provided the detection of acoustic emission events is not changed by the change in sample volume.

Kiesewetter and Schiller[95] have shown for aluminum that V^2_{rms} is directly proportional to specimen length (no measurements with respect to sample cross-sectional area were reported). James and Carpenter[153] have investigated the effects of sample size on the acoustic emission from 7075 aluminum during tensile deformation. A large number of samples of various sizes were pulled to the same strain, and the total counts and integral of V_{rms} measured. The sum of counts was directly proportional to gage volume, while the integral V_{rms} was proportional to the square root of the gage volume.

If the distance over which the source operates is not small compared to the sample cross-section, then the number of events will no longer be proportional to sample volume. For single crystals where the acoustic emission events were associated with formation of a slip band which traversed the entire width of the sample, Fisher and Lally found that the event (burst) rate was proportional to sample length and independent of cross-sectional area.[56]

Kim *et al.*[154] recently studied the effect of specimen size on the acoustic emission from two NaCl single crystals. They took great care that they had the same deformation behavior irrespective of sample size. The amplitude distribution of the acoustic emission pulses from the two samples was measured and the relation between the number of pulses N_i and their amplitude V_i was found to be

$$N_i = A\,V_i^{-m} \tag{18}$$

where A and m are constants. The value of m was found to decrease slightly with increasing strain and the magnitude of m when compared at the same strain was smaller for the larger specimen. Thus both the average pulses and the biggest pulses were larger for the bigger specimen. Kim *et al.* also make a valid point concerning previous investigations of the relation between acoustic emission and sample size. When the sum of counts or integral of the rms output over a wide range of strains is used, the data should be corrected for changes in strain rate during the test. The acoustic emission as a function of sample size reported by James and Carpenter[153] for polycrystalline 7075 aluminum was measured well past yield (at about 2-3 percent strain) where the strain rate was not changing rapidly. Correction for strain rate changes would not significantly alter the result.

Finally, the detection of events is not necessarily independent of sample volume. Heiple and Adams[51] showed that, for the same size source event, the detected acoustic emission energy for a beryllium compression sample (3.8 cm × 1.3 cm × 1.3 cm) was ten times smaller than for a tensile sample (0.5 cm × 0.3 cm × 2.9 cm gage). The events in this test were bubbles bursting on one surface of the sample and were formed by reaction of dilute hydrofluoric acid with the beryllium. The reduction in apparent size of the acoustic emission events in the compression samples caused the smaller events to be lost in the noise and reduced the measured burst rate during deformation.

4.11 Stress State

Hamstad and coworkers[155,156] have recently shown that the stress state significantly alters the acoustic emission from a given material. In tests on 21-6-9 and 304 stainless steels and 7075 aluminum, more acoustic emission was detected and it occurred at lower strains in uniform biaxial loading, compared to uniaxial loading. The difference was attributed to enhanced fracture of inclusions in the biaxial, as compared to the uniaxial, stress state. Whether complex stress states alter acoustic emission from dislocation motion is not known.

5. SUMMARY AND CONCLUSIONS

The evidence is overwhelming that dislocation motion creates acoustic emission during deformation of many materials. Furthermore, it is unlikely that the uncoordinated motion of individual dislocations can presently be detected by acoustic emission measurements, with the possible exception of measurements during deformation of high purity single crystals and very large-grained polycrystals. The nearly simultaneous motion of many dislocations within a small volume of material is generally required. In addition to the formation of a packet or avalanche of moving dislocations, the packet must glide far enough at a high enough velocity or the elastic waves produced will not be large enough to be detected. These conditions are not often satisfied simultaneously, so only a small fraction of the dislocation motion within a sample produces acoustic emission. Microstructural and testing variables can change the probability of avalanche formation, the glide distance, or the glide velocity to varying degrees and produce large changes in the observed acoustic emission. Equipment limitations, including particularly the sensitivity of the sensor and the frequency bandwidth over which the measurement is made, also substantially influence the acoustic emission observed.

A major problem in some situations is determining whether the acoustic emission observed in a particular test is even related to dislocation motion. Resolution of this question is aided by comparison of acoustic emission from different deformation modes; in tension and compression for example, and by careful microstructural characterization. Simultaneous measurement of another parameter sensitive to dislocation processes, such as internal friction, may also be helpful. Failure to take these precautions has resulted in misidentification of acoustic emission source mechanisms, most notably in 7075 aluminum beyond yield.

The acoustic emission observed is strongly dependent on

test parameters and metallurgical condition of the material. Increasing the strain rate increases the rate at which acoustic emission events occur within the sample, generally leading to a proportionality between the rms acoustic emission voltage and the square root of the plastic strain rate. Under some conditions, the acoustic emission mechanisms are altered by strain rate changes and the rms voltage observed is no longer porportional to the square root of plastic strain rate. Portevin-LeChatelier yielding and the accompanying acoustic emission are very strain rate sensitive. Test temperature appears to alter significantly the amount of acoustic emission observed even in the absence of an obvious change in the nature of the source mechanism. Changes in acoustic emission with test temperature can also result from changes in the source mechanism in a manner analogous to effects observed with strain rate changes. Event rate also appears to be proportional to sample volume, provided the material is polycrystalline and the grain size is small compared to the sample dimensions. The event rate or rms signal actually detected may depend in a more complicated way on sample volume if transmission of the signal from the source to the transducer is significantly altered by changes in sample size.

Cold or warm working prior to acoustic emission testing generally reduces the observed emission. The reduction probably occurs either because sources are activated which would have otherwise operated during the test, because the dislocation density is increased and thereby the glide distance for the dislocation packets is reduced, or both. Prior thermal history incluences the size, distribution, and morphology of precipitates, solute atoms, inclusions, and intentional second phases. It also affects grain size and dislocation density. All of these factors are known to influence the observed acoustic emission.

Solute atoms may serve as breakable pins (as is likely to occur during Portevin-LeChatelier yielding and at some yield points), thereby favor the formation of dislocation avalanches, and increase acoustic emission. Under other conditions, solute atoms may reduce dislocation velocity (or possibly acceleration) and thereby reduce acoustic emission. Precipitates formed during aging have been shown to increase the acoustic emission during subsequent deformation in some alloy systems and decrease it in others. A decrease in acoustic emission because of aging precipitates has generally been attributed to a reduction in the glide distance of the dislocation packets. An increase in acoustic emission in the presence of shearable precipitates has been attributed to the precipitates serving as breakable pins and increasing the probability of dislocation avalanches, or to the precipitates causing strain localization and a long glide distance in the resulting slip bands.

The effect of changing grain size on acoustic emission remains an area of controversy. It appears that the acoustic emission has a maximum at a particular grain size (which may be a single crystal), but the grain size at which the maximum occurs depends on sample purity, the frequency bandwidth over which measurements are made, and probably other factors.

Substantial progress has been made in the last few years in providing quantitative estimates of the surface displacements expected from dislocation motion. These provide valuable insight into the effects of various testing and microstructural variables on the acoustic emission produced by dislocation sources. Sensors have been developed which can measure actual surface displacements from simulated acoustic emission events and the more energetic actual acoustic emission events, but as yet these sensors are not sensitive enough to detect acoustic emission from dislocation motion except in a few unusual cases. A piezoelectric sensor with high sensitivity which accurately measures surface displacements has recently been invented which may help in the investigation of lower level acoustic emission from dislocation motion.

Understanding of acoustic emission from dislocation motion has advanced significantly over the past several years. Difficult problems remain in a quantitative description of dislocation source mechanisms and in an understanding of the precise way in which microstructural variables affect dislocation acoustic emission sources. Recent work has, however, greatly narrowed the range of reasonable explanations for the effects of material variables on acoustic emission.

6. ACKNOWLEDGMENTS

Portions of this work were supported by the Department of Energy, Albuquerque Operations Office (Contract DE-AC04-76DP03533) and by the National Science Foundation, Division of Materials Research, Grant No. DMR-1805112. Special thanks are extended to T. F. Drouillard for the use of the index and source documents from his acoustic emission bibliographies.[157,158]

REFERENCES

1. S. H. Carpenter and C. R. Heiple, Fundamentals of Acoustic Emission, K. Ono, editor (University of California, Los Angeles, 1979), pp. 49-104.
2. A. E. Lord, Physical Acoustics, W. P. Mason and R. N. Thurston, editors (Academic Press, New York, 1981), Vol. 15, pp. 295-360.
3. H. N. G. Wadley, C. B. Scruby and G. H. Speake, Int. Met. Rev. *25*, 41 (1980).
4. D. Jaffrey, Australasian Corrosion Engineering *23* (6); 9 (part 1); *23* (7-8) 9 (part 2); *23* (9-10), 19 (part 3) (1979).
5. K. Ono, Fundamentals of Acoustic Emission, K. Ono, editor (University of California, Los Angeles, 1979), pp. 167-207.
6. H. Hatano, J. Acoust. Soc. Am. *57*, 639 (1975).
7. J. S. Bendat and A. G. Piersol, Measurement and Analysis of Random Data (J. Wiley and Sons, New York, 1966), pp. 19–22, 74-76.
8. F. R. Breckenridge and M. Greenspan, J. Acoust. Soc. Am. *69*, 1177 (1981).
9. C. B. Scruby, J. C. Collingwood and H. N. G. Wadley, J. Phys. D *11*, 2359 (1978).
10. C. H. Palmer and R. E. Green, Jr., Appl. Opt. *16*, 2333 (1977).

11. C. B. Scruby, C. Jones, J. M. Titchmarsh and H. N. G. Wadley, Met. Sci. *15*, 241 (1981).
12. H. N. G. Wadley and C. B. Scruby, Advances in Acoustic Emission, H. L. Dunegan and W. F. Hartman, editors, (Dunhart Publishers, Knoxville, TN, 1981) pp. 125-153.
13. H. N. G. Wadley and C. B. Scruby, Acta Metall. *27*, 613 (1979).
14. C. B. Scruby, H. N. G. Wadley, K. Rusbridge and D. Stockham – Jones, Met. Sci. *15*, 599 (1981).
15. T. Imanaka, K. Sano, and M. Shimizu, Bull. Jpn. Inst. Met. *12*, 871 (1973). In Japanese, English translation UCRL-TRANS-10747 from NTIS.
16. A. B. L. Agarwal, J. R. Frederick, and D. K. Felbeck, Metall. Trans. *1*, 1069 (1970).
17. C. A. Tatro, Acoustic Emission, ASTM STP 505 (American Society for Testing and Materials, Philadelphia, PA, 1972), pp. 84-99.
18. C. B. Scruby, H. N. G. Wadley, R. J. Dewhurst, D. A. Hutchins, and S. B. Palmer, AERE-R10003 (United Kingdom Atomic Energy Authority, Harwell, U.K., 1981).
19. T. M. Proctor, Jr., J. Acoustic Emission *1*, 173 (1982).
20. D. O. Harris, A. S. Tetelman, and F. A. Darwish, Acoustic Emission, ASTM STP 505 (American Society for Testing and Materials, Philadelphia, PA, 1972), pp. 243-246.
21. E. Esmail and I. Grabec, Ultrasonics *16*, 87 (1978).
22. D. M. Egle, J. Acoust. Soc. Am. *65*, 1198 (1979).
23. M. A. Hamstad and A. K. Mukherjee, UCRL-76077 (Lawrence Livermore National Laboratory, Livermore, CA, 1974).
24. K. Ono, Mater, Eval. *34*, 177 (1976).
25. M. A. Hamstad, UCRL 76286 (Lawrence Livermore National Laboratory, Livermore, CA, 1974).
26. M. A. Hamstad and A. K. Mukherjee, Exp. Mech. *14*, 33 (1974).
27. B. J. Brindley, J. Holt, and I. G. Palmer, Non-Destr. Test. *6*, 299 (1973).
28. M. A. Hamstad and R. G. Patterson, UCRL-79356 (Lawrence Livermore National Laboratory, Livermore, CA, 1977).
29. J. R. Wadin, unpublished research (Dunegan/Endevco Corp., San Juan Capistrano, CA, 1978).
30. R. W. B. Stephens and A. A. Pollock, J. Acoust, Soc. Am. *50*, 904 (1971).
31. L. J. Graham and G. A. Alers, Mater. Eval. *32*, 31 (1974).
32. B. Woodward and R. W. Harris, Acustica *37*, 190 (1977).
33. A. G. Beattie, Fundamentals of Acoustic Emission, K. Ono, editor (University of California, Los Angeles, 1979), pp. 17-47.

34. D. M. Egle, C. A. Tatro, and A. E. Brown, Mater. Eval. *39*, 1037 (1981).
35. S. H. Carpenter and F. P. Higgins, Proceedings of the Third Acoustic Emission Symposium (Japan Industrial Planning Association, Tokyo, Japan, 1976), pp. 288-304.
36. N. N. Hsu, J. A. Simmons and S. C. Hardy, Mater. Eval. *35*, 100 (1977).
37. F. P. Higgins and S. H. Carpenter, Acta Metall. *26*, 133 (1978).
38. T. Imanaka, K. Sano, and M. Shimizu, Crystal Lattice Defects *4*, 57 (1973).
39. J. C. Duke, Jr., and R. E. Green, Jr., Proceedings of the Second International Conference on Mechanical Behavior of Materials, Boston MA (American Society for Metals, Metals Park, OH, 1976), pp. 1646-1650.
40. C. E. Coleman and J. F. R. Ambler, Zirconium in the Nuclear Industry, ASTM STP 633, A. L. Lowe, Jr., and G. N. Parry, editors (American Society for Testing and Materials, Philadelphia, PA, 1977), pp. 589-607.
41. K. Okajima and K. Ono, Proceedings of the Fifth International Acoustic Emission Symposium (Japanese Society for Nondestructive Inspection, Tokyo, Japan, 1980), pp. 270-281.
42. K. Ono, M. Shibata, and M. A. Hamstad, Metall. Trans. A *10A*, 761 (1979).
43. K. Ono and M. Yamamoto, Mater. Sci. & Eng. *47*, 247 (1981).
44. D. Jaffrey, Proceedings of the Fifth International Acoustic Emission Symposium (Japanese Society for Nondestructive Inspection, Tokyo, Japan, 1980), pp. 249-260.
45. J. Holt and D. J. Goddard, Mater. Sci. & Eng. *44*, 251 (1980).
46. S. L. McBride, J. W. MacLachlan and B. P. Paradis, J. Nondestr. Eval. *2*, 35 (1981).
47. S. H. Carpenter and F. P. Higgins, Metall. Trans. A *8A*, 1629 (1977).
48. M. A. Hamstad and A. K. Mukherjee, UCRL-77502 (Lawrence Livermore National Laboratory, Livermore, CA, 1975).
49. S. McK. Cousland and C. M. Scala, Met. Sci. *15*, 609 (1981).
50. R. I. Mints, V. S. Kortov and V. P. Nelekhin, Metallofizika, Kiev 44, 93 (1973). In Russian, English translation available as RTS-9680, British Library, Lending Division, Boston Spa, Wetherby, West Yorks, England (1975).
51. C. R. Heiple and R. O. Adams, Proceedings of the Third Acoustic Emission Symposium (Japan Industrial Planning Association, Tokyo, Japan, 1976), pp. 342-359.
52. J. E. Sinclair, J. Phys. D *12*, 1309 (1979).

53. C. Scruby, H. Wadley and J. E. Sinclair, Philos. Mag. A *44*, 249 (1981).
54. J. A. Simmons and R. B. Clough, Dislocation Modeling of Physical Systems (Pergamon Press, Oxford, 1981), pp. 464-497.
55. P. P. Gillis and M. A. Hamstad, Mater. Sci. & Eng. *14*, 103 (1974).
56. R. M. Fisher and J. S. Lally, Can. J. Phys. *45*, 1147 (1967).
57. C. R. Heiple, unpublished research (Rockwell International, Rocky Flats Plant, Golden, CO).
58. D. R. James and S. H. Carpenter, J. Appl. Phys. *42*, 4685 (1971).
59. T. Kishi and K. Kuribayashi, Fundamentals of Acoustic Emission, K. Ono, editor (University of California, Los Angeles, 1979), pp. 105-128.
60. K. Kuribayashi and T. Kishi, Acoustic Emission (Deutche Gesellschaft fur Metallkunde, Germany, 1980), pp. 53-60.
61. K. Kuribayashi, T. Kishi, K. Ohno, and R. Horiuchi, Advances in Acoustic Emission, H. L. Dunegan, and W. F. Hartman, editors (Dunhart Publishers, Knoxville, TN, 1981), pp. 112-124.
62. P. Fleischmann, F. Lakestani, and J. C. Baboux, Mater. Sci. & Eng. *29*, 205 (1977).
63. D. Rouby and P. Fleischmann, Phys. Status Solidi A *48*, 439 (1978).
64. K. Kuribayashi and T. Kishi, Mater. Sci. & Eng. *33*, 159 (1978).
65. J. J. Gilman, J. Appl. Phys. *36*, 2772 (1965).
66. A. S. Tetelman, Proceedings of the U.S.-Japan Joint Symposium on Acoustic Emission (Japan Industrial Planning Association, Tokyo, Japan, 1972), English volume, pp. 1-46.
67. N. L. Johnson and S. Kotz, Continuous Univariate Distributions – 1 (John Wiley and Sons, New York, 1970), pp. 250-271.
68. C. R. Heiple, Paper Summaries, ASNT Fall Conference, Oct. 2-5, 1978, Denver, CO (American Society for Nondestructive Testing, Columbus, OH, 1978), pp. 113-118.
69. J. Baram and M. Rosen, Mater. Sci. & Eng. *40*, 21 (1979).
70. W. F. Hartman, C00-2204-14 (Johns Hopkins University, Baltimore, MD, 1975).
71. W. F. Hartman, Exp. Mech. *14*, 19 (1974).
72. H. Hatano, J. Appl. Phys. *47*, 3873 (1976).
73. J. Holt, I. G. Palmer and D. J. Goddard, The Detection and Measurement of Cracks (The Welding Institute, Abington Hall, Cambridge, England, 1976), pp. 22-27.
74. S.-Y. S. Hsu and K. Ono, Proceedings of the Fifth International Acoustic Emission Symposium (Japanese Society for Nondestructive Inspection, Tokyo, Japan, 1980), pp. 283-293.

75. S.-Y. S. Hsu and K. Ono, Proceedings of the Fifth International Acoustic Emission Symposium (Japanese Society for Nondestructive Inspection, Tokyo, Japan, 1980), pp. 294-303.
76. S.-Y. S. Hsu, K. Ono, and H. Hatano, Mater. Sci. & Eng. *38*, 187 (1979).
77. R. Pascual, Scr. Metall. *8*, 1461 (1974).
78. T. Takagaki and A. K. Mukherjee, Strength of Materials and Alloys, P. Haasen, V. Gerold, and G. Kostorz editors (Pergamon Press, Oxford, England, 1979), Vol. 1, pp. 589-593.
79. G. T. Hahn, Acta Metall. *10*, 727 (1962).
80. D. R. Floyd, RFP-2061, (Rockwell International, Rocky Flats Plant, Golden, CO, 1974).
81. A. Granato and K. Lucke, J. Appl. Phys. *27*, 583 (1956).
82. A. Granato and K. Lucke, J. Appl. Phys. *27*, 789 (1956).
83. C. R. Heiple and S. H. Carpenter, Paper Summaries, ASNT National Spring Conference, April 2-5, 1979, San Diego, CA (American Society for Nondestructive Testing, Columbus, OH, 1979), pp. 23-28.
84. H. Nakasa, Nuclear Power Plant Control and Instrumentation STI/PUB/341, (International Atomic Agency, Vienna, Austria, 1973), pp. 461-478.
85. H. Nakasa, Y. Tomoda, and H. Takaku, Report 73027 (Central Research Institute of Electric Power Industry, Technical Laboratory No. 1, Tokyo, Japan, 1973).
86. C. R. Heiple, S. H. Carpenter, and M. J. Carr, Met. Sci. *15*, 587 (1981).
87. J. R. Frederick, AFML-TR-72-114 (Air Force Materials Laboratory, Wright-Patterson Air Force Base, OH, 1972).
88. F. R. Breckenridge, C. E. Tschiegg, and M. Greenspan, J. Acoust. Soc. Am. *57*, 626 (1975).
89. N. N. Hsu, J. A. Simmons, and S. C. Hardy, Mater. Eval. *35*, 100 (1977).
90. J. E. Michaels, T. E. Michaels, and N. Sachse, Mater. Eval. *39*, 1032 (1981).
91. K. Malen and L. Bolin, Phys. Status Solidi (B) *61*, 637 (1974).
92. K. N. Tandon and K. Tangri, Mater. Sci. & Eng. *20*, 47 (1975).
93. C. K. Day, M. S. Thesis (Washington State University, Pullman, WA, 1969).
94. V. D. Natsik and K. A. Chishko, Sov. Phys.-Solid State *14*, 2678 (1973).
95. N. Kiesewetter and P. Schiller, Phys. Status Solidi (A) *38*, 569 (1976).
96. A. Ookawa and K. Yazu, J. Phys. Soc. Jap. *18* (Supplement I), 36 (1963).

97. B. Cortellazzi, S. Boffi, G. Caglioti, and F. Rossitto, J. Appl. Phys. *44*, 1518 (1973).
98. V. D. Natsik and A. N. Burkanov, Sov. Phys.-Solid State *14*, 1111 (1972).
99. V. D. Natsik and K. A. Chishko, Sov. Phys.-Solid State *20*, 1117 (1978).
100. H. Hatano, H. Tanaka, R. Horiuchi, and N. Niwa, J. Jap. Inst. Met. *39*, 675 (1975).
101. T. Kishi, Y. Sakakibara, H. Tanaka, R. Horiuchi, and Y. Obata, Proceedings of the Third Acoustic Emission Symposium (Japan Industrial Planning Association, Tokyo, Japan, 1976), pp. 305-325.
102. G. G. Martin, Proceedings of the Ninth World Conference on Non-Destructive Testing (Australasian Institute of Metals, Parkville, Victoria, Australia, 1979), Paper 4J-4.
103. H. Imaeda, H. Kusanagi, H. Kimura, and H. Nakasa, Proceedings of the Third Acoustic Emission Symposium (Japan Industrial Planning Association, Tokyo, Japan, 1976), pp. 492-512.
104. M. Shimizu, T. Imanaka, and K. Fujimoto, Proceedings of the Third Acoustic Emission Symposium (Japan Industrial Planning Association, Tokyo, Japan, 1976), pp. 369-392.
105. K. Sano, T. Imanaka, T. Funakoshi, and K. Fujimoto, Proceedings of the Second Acoustic Emission Symposium (Japan Industrial Planning Association, Tokyo, Japan, 1974), Session 5, pp. 23-45.
106. P. Fenici, N. Kiesewetter, and P. Schiller, EUR-5550, e (Commission on the European Communities, Biannual Report 1974-1975, Luxembourg, 1976) pp. 173-177.
107. M. Friesel, Doctoral Thesis (Denver University, Denver, CO, 1982).
108. E. R. Naimon, H. M. Ledbetter, and W. F. Weston, J. Mater. Sci. *10*, 1309 (1975).
109. H. Hatano, J. Appl. Phys. *48*, 4397 (1977).
110. A. T. Green and H. L. Dunegan, Proceedings of the First International Conference on Structural Mechanics in Reactor Technology (EUR-4820, Commission of the European Communities, Brussels, Belgium, 1972), Vol. 4, pp. 507-527.
111. S. P. Ying, Crit. Rev. Solid State Sci. *4*, 85 (1973).
112. T. E. Michaels and H. M. Fraser, BNWL-545 (Battelle Pacific Northwest Laboratory, Richland, WA, 1967).
113. S. H. Carpenter, Unpublished Research (Denver University, Denver, CO).

114. S. H. Carpenter and A. Hadjicostis, Instrumentation in the Aerospace Industry, Vol. 24, and Advances in Test Measurement, Vol. 15, Part 2, K. E. Kissell, editor, (Instrument Society of America, Pittsburgh, PA, 1978), pp. 445-450.
115. R. T. Sedgwick, J. Appl. Phys. *39*, 1728 (1968).
116. H. L. Dunegan and C. A. Tatro, Techniques of Metals Research, R. F. Bunshah, editor (Interscience Publishers, New York, 1971), Vol. 5, Part 2, pp. 273-312.
117. J. H. Williams and G. P. Emerson, Mater. Eval. *37* (11), 74 (1979).
118. I. I. Averbukh and V. E. Vainberg, Soviet J. Nondestr. Test *9*, 401 (1973).
119. T. Kishi, Y. Sakakibara, H. Tanaka, R. Horiuchi, and Y. Obata, Proceedings of the Third Acoustic Emission Symposium (Japan Industrial Planning Association, Tokyo, Japan, 1976), pp. 305-325.
120. T. Kishi, T. Tanaka, and R. Horiuchi, Scr. Metall. *9*, 1023 (1975).
121. A. Hadjicostis and S. H. Carpenter, Paper Summaries, ASNT Fall Conference, Oct. 2-5, 1978, Denver, CO (American Society for Nondestructive Testing, Columbus, OH, 1978), pp. 380-384.
122. H. Borchers and H. M. Tensi, Z. Metallk. *53*, 692 (1962).
123. R. W. Armstrong and N. R. Borch, Metall. Trans. *2*, 3073 (1971).
124. Kh. G. Schmitt-Thomas, H. M. Tensi, and H. Zeitler, Aluminium *51*, 520 (1975). In German, English translation in the English supplement to the journal.
125. Kh. G. Schmitt-Thomas, H. M. Tensi, and H. Zeitler, Aluminium *53*, 719 (1977). In German, English translation in the English supplement to the journal.
126. J. Masuda, Proceedings of the Fifth International Acoustic Emission Symposium (Japanese Society for Nondestructive Inspection, Tokyo, Japan, 1980), pp. 304-317.
127. J. Masuda, Proceedings of the Fourth Acoustic Emission Symposium (International Technical Exchange Center, Tokyo, Japan, 1978), pp. 2-19 to 2-31.
128. V. A. Phillips, Trans. Metall. Soc. AIME *230*, 967 (1964).
129. D. E. Laughlin and J. W. Cahn, Acta Metall. *23*, 329 (1975).
130. J. A. Cornie, A. Datta, and W. A. Soffa, Metall. Trans. *4*, 727 (1973).
131. Kh. G. Schmitt-Thomas, H. M. Tensi, and H. Zeitler, Acoustic Emission (Deutche Gesellschaft fur Metallkunde, Germany, 1980), pp. 17-38.
132. G. Green, Met. Sci. *15*, 505 (1981).

133. R. J. Landy and K. Ono, J. Acoustic Emission *1*, 7 (1982).
134. G. V. Rao and R. Gopal, Periodic Inspection of Pressurized Components (Mechanical Engineering Publications Ltd., London, England, 1979), pp. 275-293.
135. H. N. G. Wadley, C. B. Scruby, P. Lane, and A. Hudson, Met. Sci. *15*, 514 (1981).
136. J. A. Gorman, D. S. Wood, and T. Vreeland, J. Appl. Phys. *40*, 833 (1969).
137. S. Mintzer, R. Pascual, and R. M. Volpi, Scr. Metall. *12*, 531 (1978).
138. M. A. Hamstad, unpublished research (Lawrence Livermore National Laboratory, Livermore, CA).
139. R. C. Bill, J. R. Frederick, and D. K. Felbeck, J. Mater. Sci. *14*, 25 (1979).
140. J. Baram and M. Rosen, Mater. Sci. & Eng. *45*, 255 (1980).
141. J. Baram and M. Rosen, Mater. Sci. & Eng. *47*, 243 (1981).
142. H. Hatano and N. Niwa, Conference Proceedings, The Seventh International Conference on Nondestructive Testing, Warsaw, Poland, June 4-8, 1973 (Instytut Podstawowych Problemow Techniki Polskiej Akademii Nauk, Warszawa, Poland, 1973), Vol. II, Paper E-12, pp. 225-229.
143. H. Hatano, N. Niwa, T. Kishi, and R. Horiuchi, Hihakai Kensa *21*, 219 (1972).
144. H. C. Kim and T. Kishi, Phys. Status Solidi (A) *55*, 189 (1979).
145. T. Nakamura, N. W. Ringshall, Y. Fukuzawa, and A. Adachi, Proceedings of the Fifth International Acoustic Emission Symposium (Japanese Society for Nondestructive Inspection, Tokyo, Japan, 1980), pp. 318-325.
146. B. H. Schofield, ASD-TDR-63-509, Part II (Lessels and Associates, Inc. Waltham, MA, 1964).
147. R. G. Liptai, Doctoral Thesis (Michigan State University, East Lansing, MI, 1963).
148. K. Kuribayashi, H. Tanaka, T. Kishi, R. Horiuchi, and A. Kato, Proceedings of the Third Acoustic Emission Symposium (Japan Industrial Planning Association, Tokyo, Japan, 1976), pp. 217-287.
149. M. Shimizu, K. Sano, T. Imanaka, and K. Fujimoto, Proceedings of the Second Acoustic Emission Symposium (Japan Industrial Planning Association, Tokyo, Japan, 1974), Session 5, pp. 1-22.
150. J. P. Toronchuk, Doctoral Thesis (University of Manitoba, Winnipeg, Manitoba, 1974).
151. F. M. Mazzolai and H. C. Kim, Journees d' Etudes sur l'Emission Acoustique, P. F. Gobin, R. Goutte, and C. Guillaud, Organizers

(Institut National des Sciences Appliques de Lyon, Lyon, France, 1975), pp. 110-118.
152. P. Jax and J. Eisenblatter, Battelle Information 15 (Battelle-Institut eV, Frankfurt am Main, Germany, 1973), pp. 2-8.
153. D. R. James and S. H. Carpenter, Scr. Metall. *10*, 779 (1976).
154. H. C. Kim, D. E. C. Elvin, and G. Rosenhouse, Philos. Mag. A. *40*, 49 (1979).
155. M. A. Hamstad, E. M. Leon, and A. K. Mukherjee, Met. Sci. *15*, 541 (1981).
156. M. A. Hamstad, R. G. Patterson, and A. K. Mukherjee, Proceedings of the Ninth World Conference on Nondestructive Testing (Australasian Institute of Metals, Parkville, Victoria, Australia, 1979), Paper 4J-10.
157. T. F. Drouillard, Acoustic Emission: A Bibliography with Abstracts, (IFI/Plenum, New York, 1979).
158. T. F. Drouillard, Supplement I to Acoustic Emission: A Bibliography with Abstracts, (IFI/Plenum, New York, publication anticipated in 1983).
159. H. N. G. Wadley, Private Communication, 1982.

Acoustic Emission Signal Characterization

Ronald A. Kline
The University of Oklahoma
School of Aerospace, Mechanical and Nuclear Engineering
865 Asp Avenue
Norman, Oklahoma 73019

1. INTRODUCTION

Acoustic emission (AE), the phenomenon of internal elastic wave generation in stressed materials, is a subject which has attracted research interest for the past 50 years. For example, tin cry, the audible sounds accompanying the deformation of tin, was studied as early as 1929.[1] Similar observations for a variety of deformation processes were made over the next 20 years.[2–4] However, these investigations were only concerned with a few materials subjected to special loading conditions. It remained for Kaiser[5] to perform the first comprehensive investigation in this area. Kaiser clearly demonstrated that internal stress wave generation was a widely occurring phenomenon with every material tested (zinc, steel, aluminum, copper, lead, wood, etc.) exhibiting the effect and that acoustic emission was intimately related to deformation. Schofield[6] and Tatro[7] expanded on the work of Kaiser in comprehensive studies of the relation between AE and deformation in a large variety of materials. Interest in internal elastic wave generation was further heightened with the first successful use of AE for structural integrity monitoring at Aerojet-General in 1964.[8] Since this time the field has grown substantially.

The primary objective of acoustic emission work, either for research or nondestructive testing purposes, has been to acquire sufficient knowledge of acoustic emission events to relate the observed elastic waves to the mechanisms responsible for their generation. Potentially, AE is a powerful technique for monitoring areas inaccessible by conventional methods. Certainly, many aspects of acoustic emission behavior are better understood in light of the great amount of recent work in the field. Data acquisition and processing techniques have achieved great sophistication in a relatively short period of time. The capability for source location in complex structures exists today. However, more fundamental research in the area AE signal analysis needs to be completed before this potential will be completely realized.

2. AE SIGNAL ANALYSIS

2.1 Ring Down Counts

Most acoustic emission signals, whether detected by accelerometer, piezoelectric transducer, capacitance microphone, electromagnetic sensor, or optical interferometer, have been found to be of relatively short duration (0 ~ 1 msec.), reach maximum amplitude early in the signal (usually on the first cycle), and decay nearly exponentially to the level of background noise (Figure 1). It is commonly assumed[9] that sensor output can be modeled as

$$V(t) = V_0 e^{-\gamma t} \sin \omega t \qquad (1)$$

where

$V(t)$ = output voltage of sensor
V_0 = initial signal amplitude
γ = decay constant (> 0)
t = time
ω = signal frequency.

Most AE signals can be classified either as burst emission (Figure 2a, signal amplitudes appreciably larger than back-

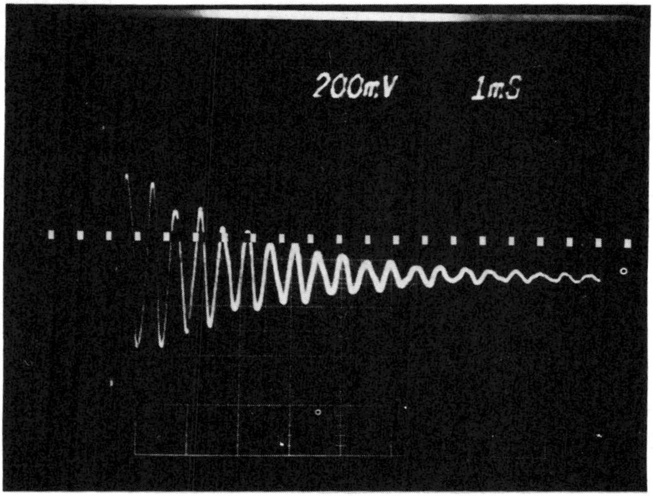

Figure 1. Simulated acoustic emission event.

ground, of short duration, and well separated in time of occurrence) or as continuous emission (Figure 2b, signal amplitudes slightly higher than background and events closely spaced in time to form a single waveform).

Most often AE signals are quantified by counting the number of times the sensor voltage exceeds a fixed threshold voltage (V^* in Figure 1). This technique is known as ring down counting. Ring down counts are usually plotted either as total counts (N) or as count rate ($\dot{N}$ = the number of counts/unit time). For the idealized signal represented by Eq. (1), the number of counts (N) to the nearest integer is given by

$$N = \frac{t^*}{2\pi/\omega} = \frac{\omega}{2\pi\gamma} \ln \frac{V_o}{V^*} \tag{2}$$

where $V^* = V_o e^{-\gamma t^*}$

and
$$t^* = \frac{1}{\gamma} \ln \frac{V_o}{V^*}. \tag{3}$$

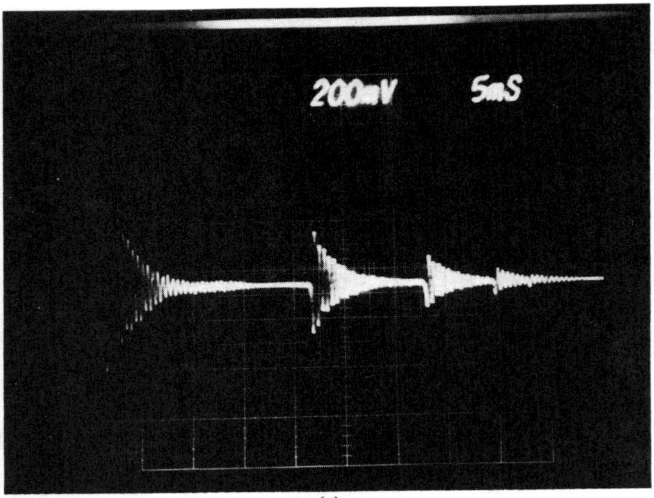

(a)

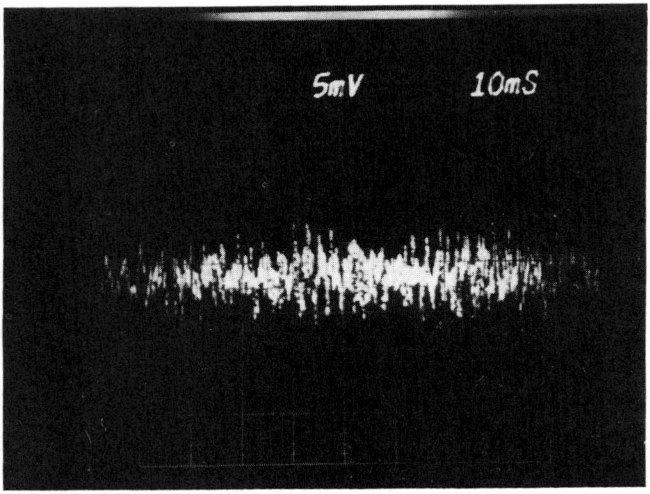

(b)

Figure 2. a) Simulated burst emission
b) Simulated continuous emission

Even for this simplified case, relating ring down counts to a single physically meaningful parameter is a difficult if not impossible task. Ring down counts are a complicated function of the frequency response of the transducer as well as the event, the decay characteristics of the transducer event and propagation media (γ), the amplitude of the signal, coupling efficiency, sensitivity of the probe, and amplifier gain (V_o); and the threshold voltage (V^*). In addition, ring down count data are sensitive to changes in amplifier gain, threshold level, and coupling efficiency (piezoelectric sensors). This is demonstrated in Figure 3 where the effect of small changes in voltage gain on acoustic emission data is shown and in Figure 4 where pulse rate as a function of threshold level for six different signal levels is presented. The importance of insuring that these factors remain constant throughout the duration of a test and be reproducible from test to test cannot be over stressed.

If ring down counts can be likened to a single parameter, it is probably closest to the energy present in an acoustic emission event. The electrical energy (E) present in a transient event is defined as

$$E = \frac{1}{R}\int_0^\infty V^2(t)\,dt \tag{4}$$

where R is the resistance of the measuring circuit.

Equations 1 and 4 can be combined to yield (assuming the signal decays to background level (V^*) after a time t^*)[11]

$$\begin{aligned} E &= \frac{1}{R}\int_0^\infty V_o^2 e^{-2\gamma t}\sin^2\omega t\,dt \\ &= \frac{V_o^2}{4R(\omega^2+\gamma^2)}\left[\frac{\omega^2}{\gamma}(1-e^{-2\gamma t^*}) - \right. \\ &\quad \left. - 2e^{-2\gamma t^*}\sin\omega t^*(\gamma\sin\omega t^* + \omega\cos\omega t^*)\right]. \end{aligned} \tag{5}$$

If it is assumed that $e^{-2\gamma t^*} << 1$, then Eq. (5) can be written as

$$E \cong \frac{V_o^2\omega^2}{4R\gamma(\gamma^2+\omega^2)}\,.$$

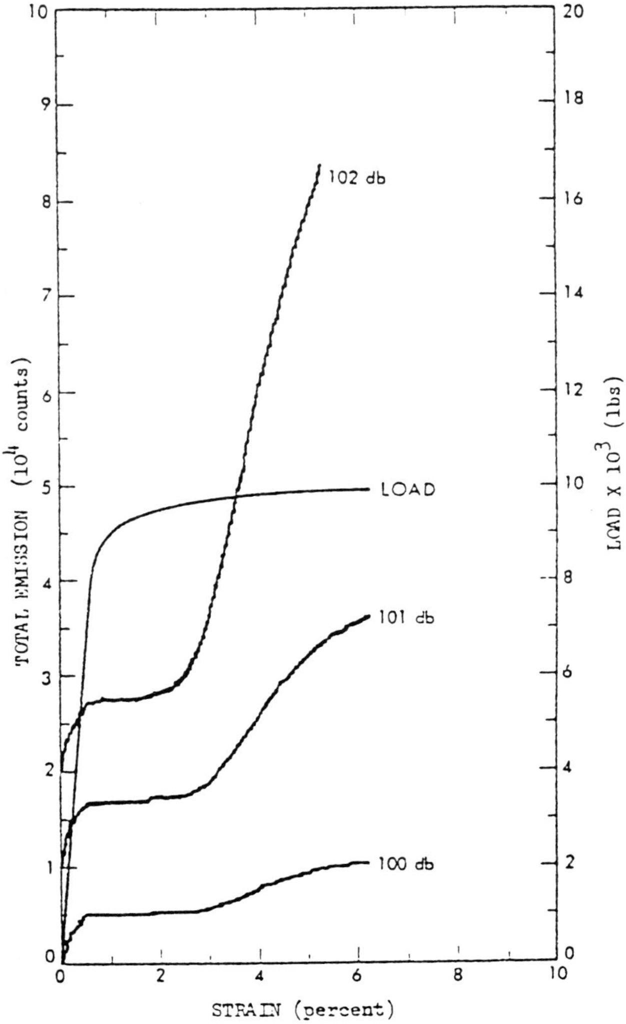

Figure 3. The effect of voltage gain on AE behavior (after Hamstad and Mukherjee). (15)

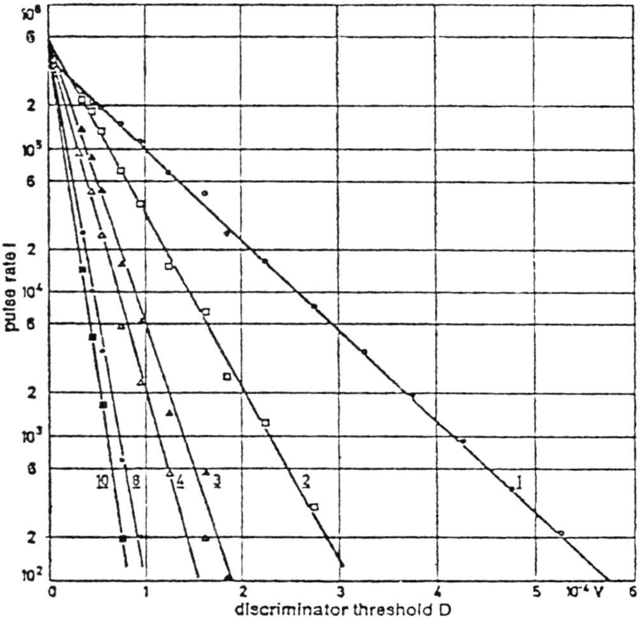

Figure 4. The effect of threshold level setting on AE count rate (after Jax and Eisenblätter). (10)

Using Eq. (2), we can write

$$V_o = V^* e^{2\pi\gamma N/\omega}. \tag{6}$$

This requires that

$$E = \frac{V^{*2}\omega^2}{4R\gamma(\gamma^2 + \omega^2)} e^{4\pi\gamma N/\omega}. \tag{7}$$

Therefore, for signals where ω = constant (either from the event itself or a narrow bandwidth sensing system), a simple exponential relationship exists between electrical energy and ring down counts. As most AE signals are wide band in nature, ring down counts provide only a qualitative measure of the energy present in acoustic emission events. For quantitative energy analysis, other techniques must be used.

2.2 Energy Analysis

Direct energy analysis can be achieved by two available methods: 1) by digitizing and integrating the waveform signal in accordance with Eq. (4), or 2) by designing a device to perform the integration electronically. As A-D conversion devices with the frequency resolution and storage capacity to completely capture an AE event are not readily available, the second alternative is usually chosen.[11, 12, 13] The advantage of this method over ring down counting is that energy measurements can be directly related to an important physical parameter, which is merely approximated as in the case of the ring down counts. If the sensor response is proportional to the displacements associated with the passage of an acoustic wave, then the electrical energy in the sensor output will be proportional to the mechanical energy in the emission event. While this is a very strong assumption for accelerometers and piezoelectric elements, it is met for the capacitance microphone, electromagnetic sensor, and optical interferometer.

Experiments conducted by Beattie[11] show that, as expected, for narrow band signals direct energy analysis yields substantially the same information as ring down counts. However, for cases where the frequency does change, differences exist between ring down and energy records. Figures 5 and 6 illustrate these differences. Both depict the energy and ring down counts observed for two hydro tests of filament wound epoxy bottles. For the test shown in Figure 5, only one type of emission event was observed and the ring down and energy curves are quite similar. For the test shown in Figure 6, however, two distinct waveforms were observed – a low energy kind which dominated the initial part of the test and a high energy type which occurred more often near failure. The two

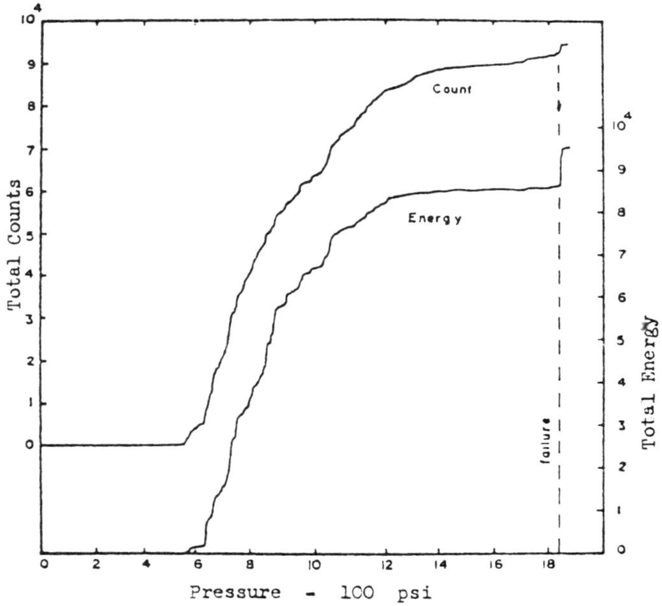

Figure 5. Comparison between count and energy analysis (after Beattie). (11)

signal types were found to have differing spectra. Notice that the initial part of the energy curve is smooth while the latter part is serrated. The count curve is completely smooth. Energy analysis was sensitive to differences in the waveform whereas ring down counts were not.

A more widely used technique related to direct energy analysis is root mean square (rms) signal averaging. For this technique, the signal is passed through a device which measures the heating power present in the waveform. As the meter response is slow in comparison with the duration of most AE signals, rms records are indicative of average acoustic power rather than the instantaneous energy measurement of the direct approach. Of course, one could plot a summation of

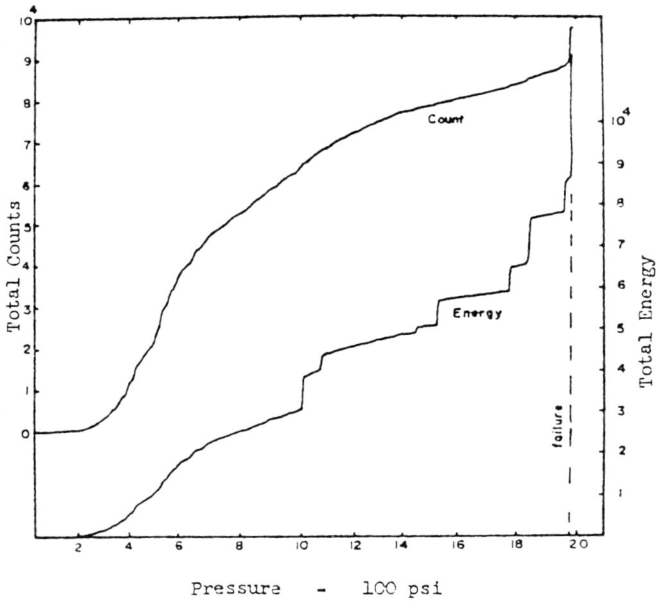

Figure 6. Comparison between count and energy analysis (after Beattie). (11)

the acoustic energy for a test which would be identical to the rms meter output. As conventionally used, substantive differences exist between rms and ring down measurements due to the averaging nature of rms readings as well as the dissimilarity between energy and rms analysis discussed in the preceding section. For most tests with ring down counters, one sets the threshold level appreciably higher than background in order to insure that only AE events are counted and not fluctuations in the background level. Thus, burst emission sensitivity is optimized at the expense of the low-level continuous emission. Conversely, rms sensitivity to transient emission and short duration noise fluctuation is relatively diminished by the averaging process. Therefore, one can measure signals closer (in amplitude) to the background level without appreciably affecting the rms output voltage appreciably.

For certain applications the differences between the two responses are quite helpful in separating various types of signals which are generated.[14] Consider the records of Figure 7 where rms and count rate are plotted as a function of time for constant strain rate deformation of a lead-tin alloy. Two distinct regions are observed. At the outset, the count activity is much larger than the rms response as the emissions observed in this region are primarily of burst type (due to microcrack formation). Later in the test, continuous emission (possibly due to slip) dominates as indicated by the large peak in the rms level while the count rate has diminished appreciably from the initial maximum. Of course, by averaging the counts over a period of time closer to that of the rms meter, the differences would be less pronounced. Hamstad and Mukherjee report that, for constant strain rate testing of 7075-T6 Al, normalizing the rms voltage output results in test reproducibility superior to that achieved for ring down counting.[15]

2.3 Event and Amplitude Distribution Analysis

Two similar rarely used methods of AE signal processing are event counting and amplitude distribution analysis. For event counting a threshold voltage is set much as for ring down counting. When an event of sufficient amplitude occurs, a trigger is fired and the counter is incremented by one. To avoid counting the same signal twice, the trigger is reset only after a time delay exceeding the ring down time of the observed signals. Obviously the technique is primarily applicable to burst emission sources. The method is an unambiguous means of counting signals but, like most automated techniques, it has a flaw. Once the trigger is fired no events which occur before the trigger is reset can be counted. For many materials and loading conditions the time between events is much larger than the decay time for AE signals. However, for a number of

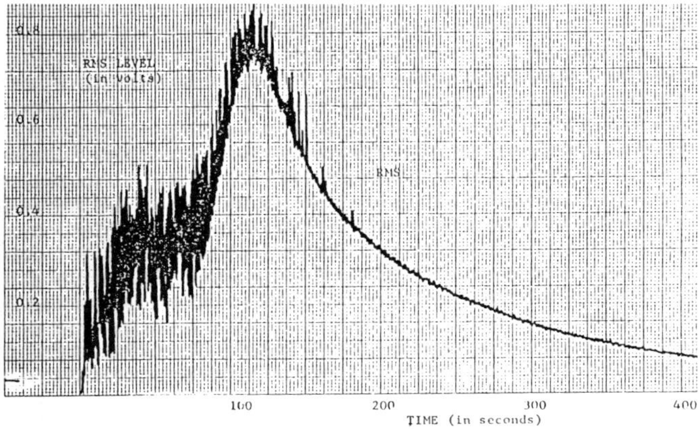

Figure 7a. Count rate vs. time for a lead-tin test with constant crosshead speed (after Kline). (14)

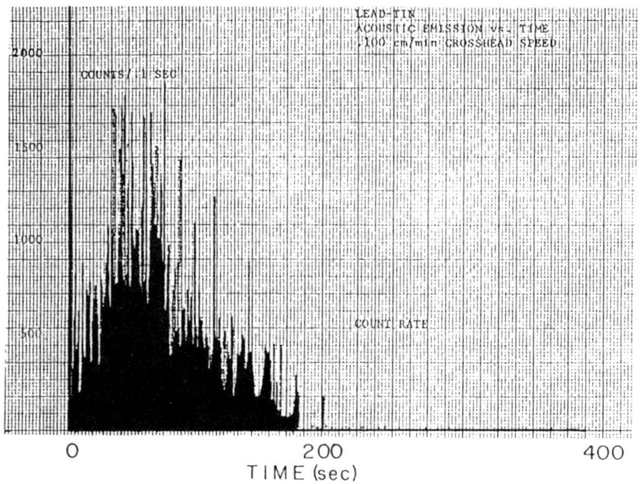

Figure 7b. RMS vs. time for the same test as in Fig. 7a (after Kline). (14)

cases the spacing between signals is relatively short and data are lost. With this method there is no way to know if the data set is complete. Amplitude distribution analysis employs a series of event counters, each set at a different threshold. By subtracting the number of events in successive registers, the number of events with amplitudes between the two threshold levels, which is the amplitude distribution, can be found. This procedure is illustrated in Figure 8 and 9 for crack growth studies of D6 AC steel of various heat treatments conducted by Nakamura *et al.*.[16] For each heat treatment the event counts at different threshold levels were determined (Figure 8). The amplitude distribution for each test was determined in the manner described above (Figure 9). In this manner the authors were able to distinguish the heat treatments producing slow crack growth (34C17, 7A35, and 25A33) from the rapid growth in the other specimens.

2.4 Correlation of AE with Other Techniques

One of the major drawbacks to any of the amplitude related signal analysis techniques described above is that, while knowledge of the magnitude and rate of emission may be suitable for nondestructive testing purposes, it yields only qualitative correlation with AE generation processes. It is not possible to quantify the complete time evolution of an acoustic emission event by a single number. Attempts[17, 18, 19] have been made to theoretically model the qualitative features of acoustic emission data with a variety of methods, mostly based on dislocation breakaway from pinning points as the generation mechanism, but success in this area has been limited. Consider the work of Dunegan and Harris[17] (Figure 10 and 11) where ring down count rate was observed for constant strain rate tensile loading of 7075-T6 Al and Fe-3% Si alloys. Attempts were made to fit these data to Gilman's expression for mobile dislocation density as a function of strain. The agreement for the aluminum specimen is excellent while the Fe-3% Si fit is

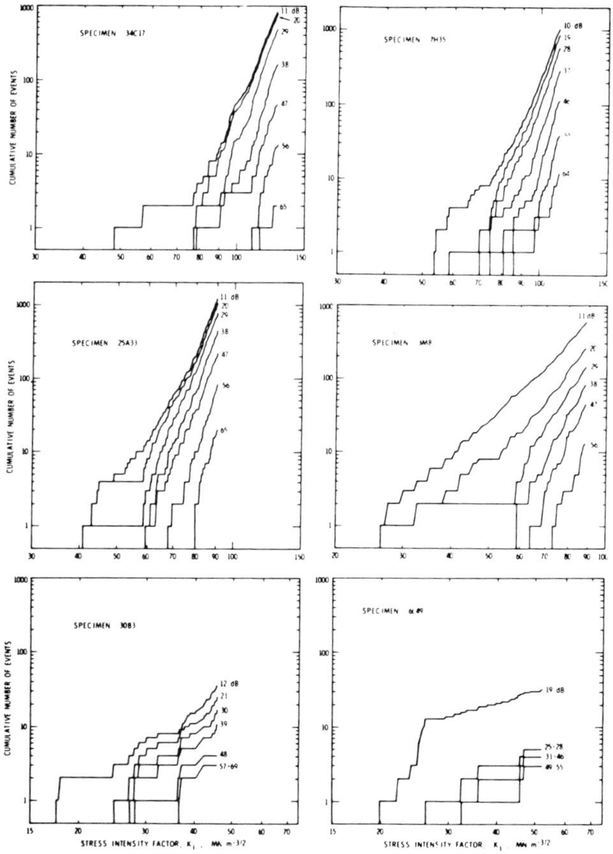

Figure 8. AE event count vs. stress intensity for a series of different threshold levels (after Nakamura *et al.*). (16)

relatively poor. In general, correlation of AE activity with any single proposed source model has been inconsistent for different materials.

Efforts have been made to experimentally correlate acoustic emission results with better understood material properties,

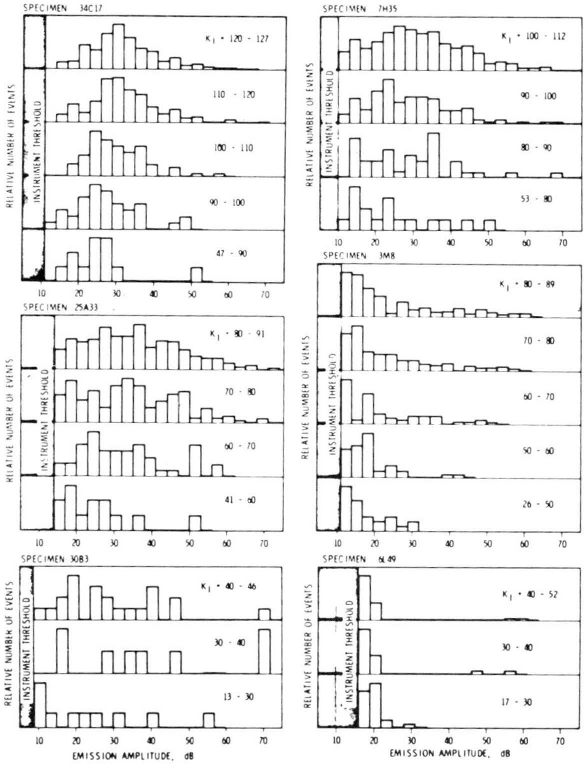

Figure 9. Amplitude distribution for the tests shown in Figure 6 (after Nakamura *et al.*). (16)

for example, ultrasonic attentuation. Imanaka *et al.*[20] compared ultrasonic attenuation of 27 MHz longitudinal waves with simultaneous AE measurements during tensile deformation of copper single crystals and polycrystalline aluminum. No obvious correlation was observed. Duke[21] monitored ultrasonic attenuation of 10 MHz longitudinal waves and acoustic emission for various aluminum alloys again with no substantive agreement. Higgins and Carpenter have measured

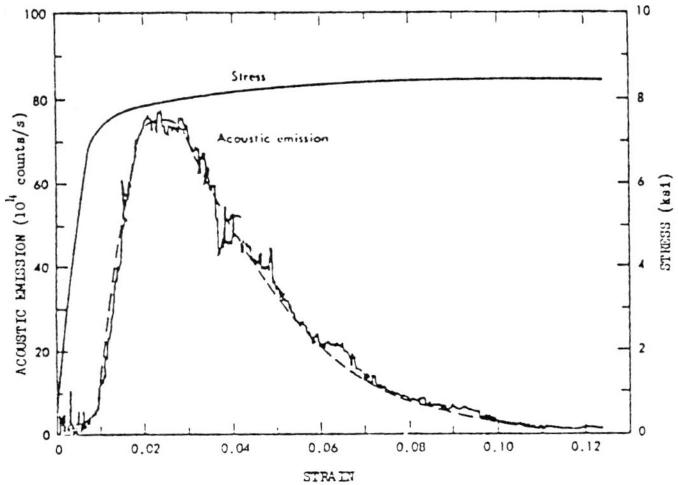

Figure 10. AE counts and stress vs. strain for constant strain rate loading of 7075-T6 Al (after Dunegan and Harris). (17)

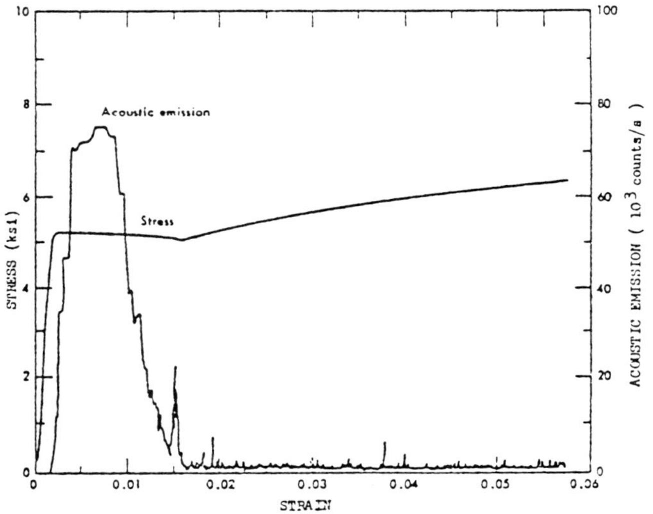

Figure 11. AE counts and stress vs. strain for constant strain rate loading of FE – 3% Si (after Dunegan and Harris). (17)

acoustic emission ring down counts and internal friction (at 50 kHz) simultaneously during constant strain rate deformation in 7075 aluminum alloys and pure iron.[22] In the 7075 aluminum, the authors report good correlation between the qualitative features of the acoustic emission and internal friction curves prior to macroscopic yield. Afterwards, the curves do not agree. The authors concluded that post yield emissions did not originate with dislocation motion. In the case of pure iron, the authors found good correlation between the acoustic emission count rate and the rate of change of the damping to indicate the probable source of acoustic emission at yield was dislocation breakaway (Figure 12). These results were compared with simultaneous magnetic strain measurements.

2.5 Frequency Analysis

Apart from amplitude related techniques the only other widely used means of examining acoustic emission signals is via spectral analysis. While early investigators realized the potential importance of frequency analysis, when both Kaiser[5] and Schofield[6] attempted to examine AE frequency behavior, they were hampered by the limited bandwidth of the available equipment. Little reliable data for the frequency range above 100 kHz exist prior to 1970. Fortunately, recent advances in electronic signal recording and processing have extended the range over which spectral analysis can accurately be performed to encompass the primary frequencies of interest in stress wave analysis. In particular, the development of the videotape recorder and high speed transient digitizer have revolutionized the AE spectral analysis field. For analog recording, the bandwidth is primarily limited by the relative velocity of the tape to the recording heads. Fixed head position machines were extremely expensive and used enormous amounts of magnetic tape to record a single test. The videotape recorder, on the other hand, employs rotating recording heads to achieve a

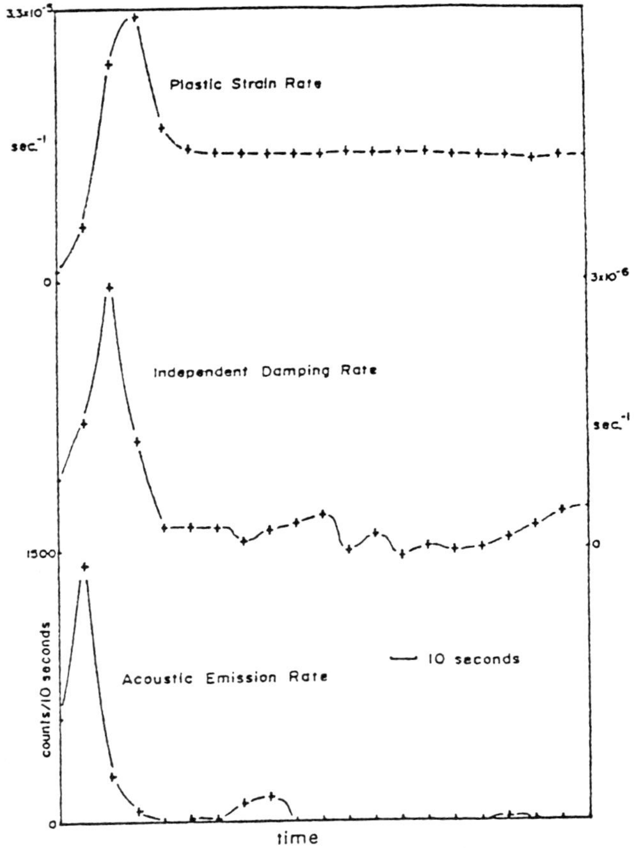

Figure 12. Plastic strain rate, independent damping and acoustic emission count rate vs. time for constant strain rate testing of iron alloy (after Higgins). (22)

bandwidth from 0-3 MHz. Also, developments in solid state electronics have resulted in analog to digital converters with sufficient high speed sampling capability and storage to capture the salient features of most AE events. The major drawbacks to the systems available are the limited dynamic range of the videotape recorder, the generation of internal synchronization signals in the videotape recorder which must be gated out to avoid biasing the spectra and the limited storage capability of the available transient digitizers which prevents complete recording of most AE signals. For example, a transient digitizer with 4096 points can only capture a .4096 msec. signal portion with 5 MHz resolution.

Two methods of spectral analysis of the recorded data are widely used today. Commercial spectrum analyzers (e.g., HP and Tektronix) are relatively inexpensive, readily available, and require only a repetitive analog signal whose repetition rate is faster than the scan rate of the analyzer, a condition easily met by the stop action mode of the videotape recorder or the analog output of the transient digitizer. Digital methods of spectral analysis, usually via the Fast Fourier Transform, are more versatile but somewhat more expensive as they require a relatively complicated interface between digitizer and computer for operation.

Interpreting the information extracted from a frequency analysis has been a challenging problem. Ono *et al.*[23] using digital techniques found no change in spectra as a function of strain for magnesium undergoing tensile deformation. Graham and Alers, using analog recording, saw no change in AE from 2024 Al, 7075 Al, 9% Ni – 4% Co steel, and 6% Al – 4% V titanium. However, during constant strain rate tensile deformation of A533-B steel, two distinct spectra associated with plastic flow and crack extension were observed.[24] Graham[25] continued this study by examining 7075-T6 Al emissions in detail. Four distinct spectra were observed, one of which he attempted to identify with the fracture of included particles in the vicinity of the fracture path. Similar attempts to explain the source have been proposed by Higgins[22] and Hamstad *et al.*.[26] No firm documentation for this claim has yet come forth. For constant strain rate testing of 1015 steel and analog processing, Bassim *et al.*[27] reported that specimens with induced flaws exhibited signals with lower frequency content than unflawed specimens. Kline,[28] for deadweight loading of yellow brass (analog processing), observed an increasing contribution of higher frequency components for emissions occurring at higher levels of strain. For HF-1 steel, a large difference in emissions occurring immediately prior to failure and prior to macroscopic yield was observed. Hatano,[29,30] using a series of resonant transducers and narrow band filters to estimate the power spectra of acoustic emission from pure aluminum and aluminum alloys, reports an increased contribution of high frequency components at higher strains as was observed in yellow brass.[28] No strain rate dependence of the power spectra was found. Ono,[31] using digital techniques, analyzed spectra from fracture toughness test of five aluminum alloys and concluded that most of the spectral features were attributable to specimen and transducer resonances.

As is readily apparent from this discussion, there is much variety in the spectral character of the acoustic emissions thus far reported. Little agreement has been reached as to whether the frequency spectra of acoustic emission events do, in fact, change as a function of deformation. In addition, the role of transducer design and specimen geometry on the frequency spectra (and the spectral changes that have been observed) has not been firmly established. It is hoped that future work will clarify this situation.

3. THE ROLE AE SENSORS IN AE CHARACTERIZATION

One of the greatest problems in acoustic emission today is that of sensor design. It has already been shown that acoustic emission techniques, whether ring down, energy, or frequency analysis, are sensitive to changes in the character of AE waveforms. However, for these measurements to be truly useful, the signals used must be accurate representations of the acoustic waves under investigation. It serves no purpose to use sophisticated processing techniques for signals which are only rough approximations to the displacements (or stress amplitudes, particle velocity, etc.) associated with the actual event. In virtually all AE tests conducted to date, the sensors fall quite short of the ideal – a colorless sensor, introducing none of its character but faithfully reproducing the signal of interest.

For acoustic emission tests, piezoelectric transducers have been the primary means of gathering AE information. While inexpensive, easy to use, and extremely sensitive, they also possess some very important drawbacks. Little is known about the accuracy of calibration of piezoelectric devices, hence little is known of the accuracy of the amplitudes associated with acoustic emission events. If one assumes a linear relationship between the output voltage and particle displacement detected for a piezoelectric element for the entire amplitude range of interest (a very strong assumption), then the typical AE sensitivity for a piezoelectric element is on the order of 10^{-15} m or (= 10^{-5} Å) as is shown below from theory developed in.[32]

$$\delta = \frac{V_{min}}{g_{33}\, E_x} \tag{8}$$

where δ = minimum detectable displacement
V_{min} = background noise level
g_{33} = piezoelectric constant
E_x = elastic modulus of transducer element

Using typical values for these parameters[32]

V_{min} = 10 microvolts
g_{33} = 24.4×10^{-3} volt·meter/Newton
E_x = 5.85×10^{10} Newton/meter^2
then δ = 7×10^{-15} meters.

Results of an independent investigation by Speake[33] indicate that this sensitivity estimate is in error by two to three orders of magnitude. In addition, it is not known whether piezoelectric devices measure solely displacement, stress amplitude, or particle velocity associated with the passage of an elastic wave or various combinations of these parameters. Recent work by Hsu *et al.*[34] shows that, at least for the sensor used, the piezoelectric output was probably more a measure of particle velocity than displacement. Sensitivity to a wide variety of waves – longitudinal, shear, Rayleigh (desirable for NDT purposes) – further complicates the precise analysis of piezoelectric measurement. Once an elastic wave enters a piezoelectric element, the sensor will reverberate until the wave is attenuated to the level of background noise. This makes true energy analysis an extremely difficult task. Also of importance, particularly for frequency analysis, is the fact that for optimal sensitivity piezoelectric transducers must be in direct contact with the surface under investigation. This applies a stress to the areas in close proximity to the element and alters the propagation of stress waves. Furthermore, the resonant nature of the transducer means that the frequency response of any piezoelectric sensor, no matter how well damped, will not be flat but will exhibit maxima and minima. This is illustrated in Figure 13 where the response of a piezoelectric transducer is compared with a sensor with flat frequency response (capacitor sensor) for broad band noise input.[35] While both show the $1/\omega^2$ frequency dependence characteristic of the event, the capacitor curve is smooth while the piezoelectric response shows numerous peaks and valleys.

Flat frequency response non-contact sensors do exist but have diminished sensitivity relative to piezoelectric elements. This is shown in Table I from Speake.[33] Due to their diminished

TABLE I
Transducer Sensitivities

Type	Limiting Sensitivity (m)
Broad Band Piezoelectric	10^{-12}
Resonant Piezoelectric	10^{-13}
Laser Interferometer	$10^{-11} - 10^{-12}$
Capacitor Microphone	10^{-12}

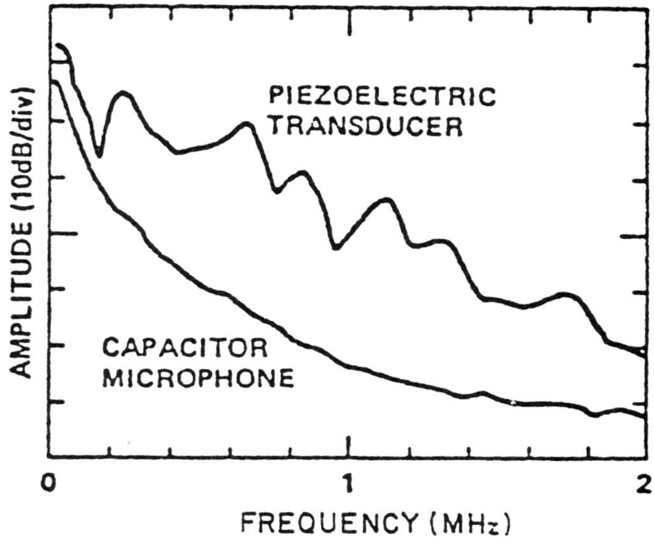

Figure 13. Response of piezoelectric and capacitance sensors to simulated white noise source (after Graham and Alers). (35)

sensitivity, flat response sensors have primarily been used for detecting large amplitude or simulated events. Capacitor microphones, which measure the capacitance changes in a DC biased electrostatic element located near the surface due to surface displacement, have been used by Graham and Alers,[35] Breckenridge *et al.*[36] and Hsu *et al.*[34] to examine simulated AE pusles. Electromagnetic sensors (estimated sensitivity 10^{-8} – 10^{-9} cm) are used to detect the eddy currents produced by the displacements accompanying the passage of an elastic wave. To date, these transducers have only been used to study events produced by martensitic phase transformation in Au – 47.5% Cd by Maxfield and Cochran.[37] A third alternative technique[38] is the use of an optical interferometer to measure changes in optical path length caused by surface displacement. Optical methods do not require direct specimen contact, hence they do not interfere with AE signal propagation.

Unlike the case of piezoelectric sensors, these alternative sensors can be precisely calibrated to allow absolute measurement of AE amplitudes. These devices measure displacement normal to the surface under investigation, whether it be due to shear, surface or longitudinal waves. The piezoelectric response to stress waves as a function of polarization and directivity is unknown. An additional advantage, particularly for the optical method, is the small sensitive area in comparison to piezoelectric techniques. This allows access to areas inaccessible by other means such as the tip of a crack.

Kline *et al.*[39] have compared piezoelectric transducer response to that of the wide band optical sensor for a variety of acoustic emission sources including stress corrosion cracking and twinning. Typical results are illustrated in Figure 14 where optical and piezoelectric response to two acoustic emission events resulting from stress corrosion cracking are presented.

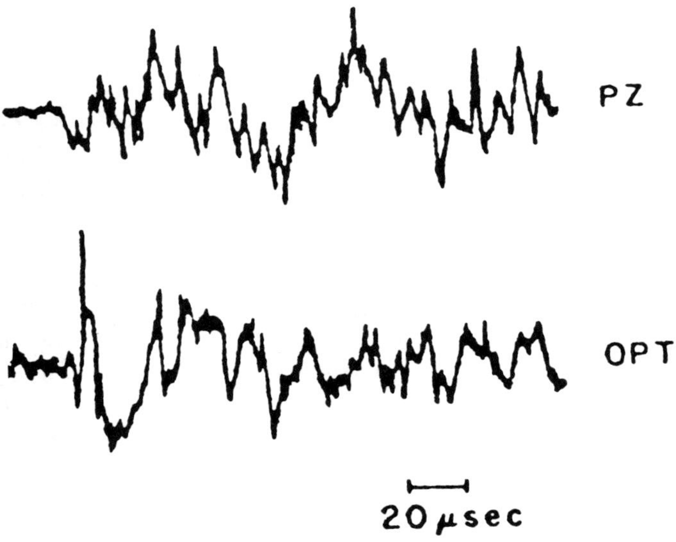

Figure 14. Comparison of optical (OPT) and piezoelectric (PZ) sensor response to acoustic emission event (after Kline *et al.*) (39)

Note that, while long term behavior is similar, the optical probe detects a large amplitude, rapid rise time "spike" at the onset of the AE signal which is undetected by the piezoelectric sensor. The importance of accurate measurement of initial surface displacements is discussed in the next section where recent advances in acoustic emission signal analysis, due in part to the development of these wide band surface displacement sensors, are presented.

4. RECENT ADVANCES IN ACOUSTIC EMISSION SIGNAL CHARACTERIZATION

The response of an elastic medium to transient loading has been studied extensively by seismologists throughout this century.[40–43] Usually, these investigations were concerned with idealized geometries and applied forces. Work in this area originates with that of Lamb[40] who studied the response of a semi-infinite elastic half space to various types of applied load including a surface point load which varied in time as a Heavyside step function. This problem is known today as Lamb's problem. Further research in the area by Caignard,[41] Knopoff,[42] and Pekeris[43] has extended these results to additional cases which have particular importance in acoustic emission applications such as epicentral response to a buried pulse in a semi-infinite medium or in a thick slab. While these studies were well known, it was not until recently that the implications of this work for acoustic emission were fully realized. Using a capacitance microphone as a wideband detector of surface displacement and the fracture of a glass capillary tube to simulate the Heavyside step function for the applied load, Breckenridge *et al.*[37] were able to demonstrate excellent agreement between the predicted displacements from seismic theory and the measured behavior of the surface for both the surface response of a simulated half space due to surface loading (Lamb's problem) as well as epicentral surface response due to a buried source. While somewhat idealized, this experiment effectively simulates, in a controlled fashion, the rapid release of elastic energy characteristic of acoustic emission events. Hsu *et al.*[34] extended this work to the case of point loading on the surface of a thick slab, a situation closely resembling that of many acoustic emission applications. A comparison of the theoretical and experimental results for the three geometries discussed above is presented in Figures 15, 16, 17 [36]

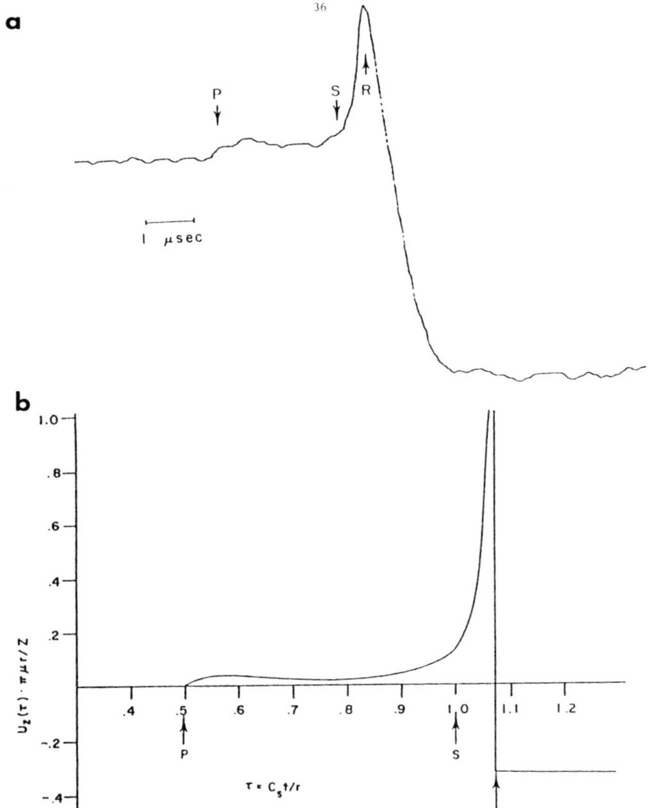

Figure 15. a) Measured surface displacement for seismic surface pulse.
b) Calculated surface displacement for seismic surface pulse (after Pekeris). (43)

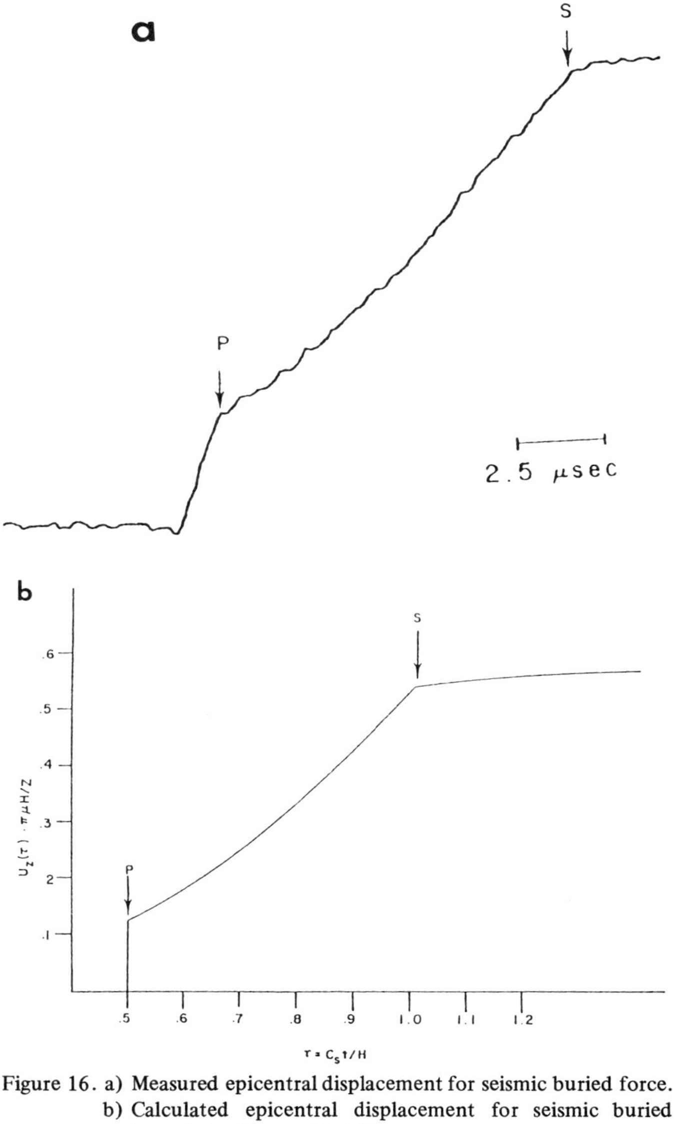

Figure 16. a) Measured epicentral displacement for seismic buried force.
b) Calculated epicentral displacement for seismic buried force (after Knopoff). (42)

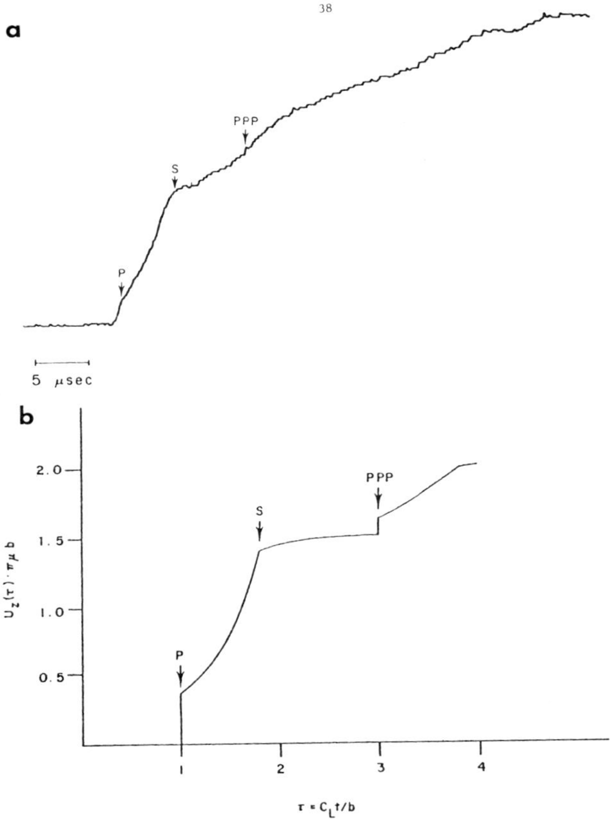

Figure 17. a) Measured epicentral displacement for thick slab.
b) Calculated epicentral displacement for thick slab (after Knopoff). (42)

where aluminum discs were used to simulate the half space and thick slab conditions, capillary tube fracture was used as the excitation source, and an optical interferometer served as a monitor of the vertical component of surface displacement.[39]

This work has led to several theoretical and experimental investigations using seismic theory as the basis. Analytically, work by Sinclair,[44] Pardee,[45] and Pao and co-workers[46,47] has greatly expanded the understanding of the response of an elastic media to transient loading to include multipole force sources, off-axis response, and more realistic source functions than the Heavyside step function assumed in the earlier investigations.

Experimentally, this approach has proven to be fruitful in the analysis of acoustic emission signals.[48, 49] In order to simulate the idealized conditions appropriate to seismic theory, careful attention must be paid to specimen geometry. In particular, efforts must be made to satisfy the conditions of the semi-infinite half space or the thick slab, and the source must be effectively isolated so that the source-receiver position is precisely known (preferably with the receiver at the epicenter). Scruby *et al.*[48] approached the problem by designing a "Yobell specimen" which localized the acoustic emission source to a small gage region but located the sensor (a capacitance transducer) in a relatively large region where the surface response to the transient load could be accurately determined before the arrival of reflected waves from the sides of the specimen. This is shown in Figure 18. Kline *et al.*[49] took a different approach. In this investigation, the thick slab geometry was simulated by a steel plate. A notch was machined in the specimen to isolate the acoustic emission source. Load was applied at the notch via a bolt and a saline solution was injected at the tip of the notch to produce stress corrosion cracking. An optical probe focussed at the tip of the crack was used to monitor the acoustic emission signals. The geometry is shown in Figure 19. In certain instances, agreement between theory and experiment is quite good as is illustrated in Figure 20 where an acoustic emission signal from the stress corrosion cracking experiment described above is compared with a theoretical solution to a buried double force located at the midplane of a thick slab. In order to more realistically simulate AE events, the theoretical solution employed a smoothly rising parabolic ramp function instead of the more commonly assumed, and more abrupt, Heavyside step function. The nearly periodic portion of the waveform is associated with reflected longitudinal waves. While these studies have enhanced our knowledge of the relationship between acoustic emission waveforms and their sources, much work remains before complete characterization of acoustic emission signals can be achieved.

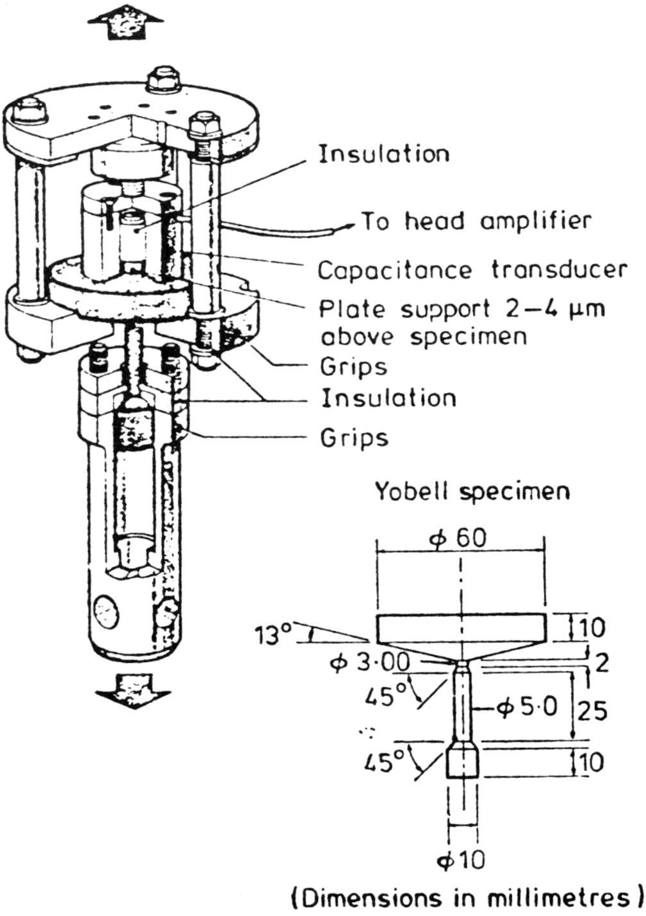

Figure 18. Yobel specimel geometry (after Scruby *et al.*). (48)

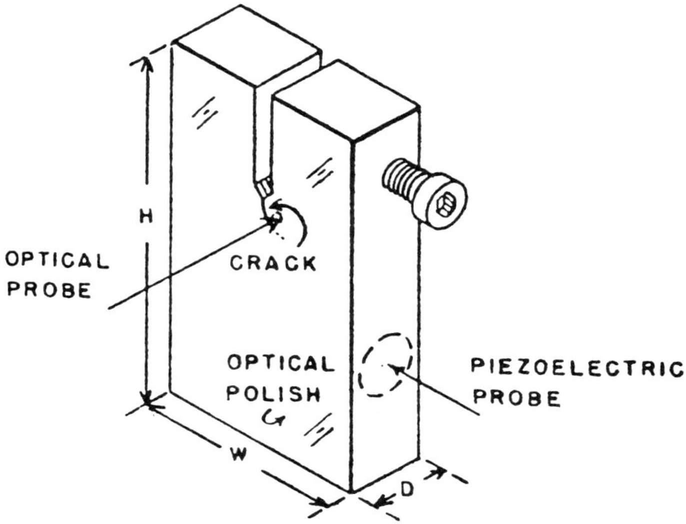

Figure 19. Stress corrosion cracking specimen geometry (after Kline *et al.*). (49)

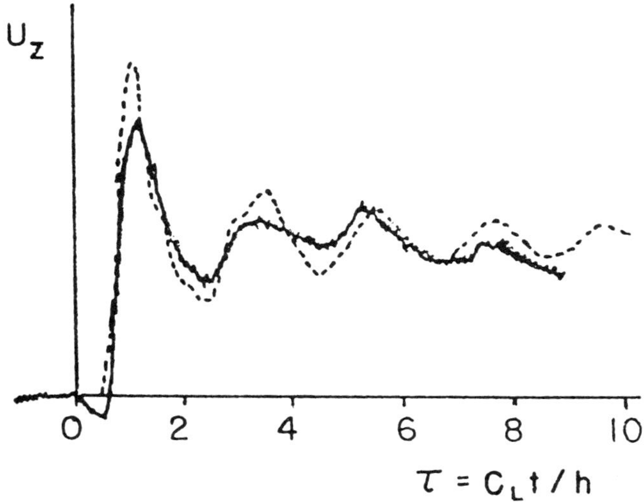

Figure 20. A comparison of predicted and optically measured acoustic emission signals (after Kline *et al.*). (49)

REFERENCES

1. M. Classen-Nekludowa, Zeitschrift für Physik, *55*, 555 (1929).
2. E. Schmid and M. Valouch, Zeitschrift für Physik, *75*, 531 (1932).
3. R. Becker and E. Orowan, Zeitschrift für Physik, *79*, 566 (1932).
4. W. Mason, H. McSkimin and W. Shockley, Physical Review, *73*, 1213 (1948).
5. J. Kaiser, Ph.D. Thesis, Techn. Hochsch., Munchen (1950).
6. B. Schofield, ASD-TDR-63-509, Lessells and Associates, Inc., Waltham, Massachusetts (1963).
7. C. Tatro, Progress Report, Michigan State University, East Lansing, Michigan (1959).
8. A. Green, C. Lockman and R. Steels, Modern Plastics, *41*, 137 (1964).
9. D. Harris, A. Tetelman and F. Darwish, ASTM STP 505, Acoustic Emission, 238, ASTM, Philadelphia (1971).
10. P. Jax and J. Eisenblätter, Information 15, Battelle Frankfurt, Frankfurt (1973).
11. A. Beattie, Materials Evaluation, *34*, 73 (1976).
12. D. Harris and R. Bell, Technical Report DE-74-3 Dunegan Endevco Corporation, San Juan Capistrano, California (1974).
13. A. Beattie and R. Jaramillo, Review of Scientific Instruments, *45*, 352 (1974).
14. R. Kline, DAAD05-75-M-1569, Johns Hopkins University, Baltimore (1976).
15. M. Hamstad and A. Mukherjee, Experimental Mechanics, *14*, 33 (1974).
16. Y. Nakamura, C. Veach and B. McCauley, ASTM STP 505, Acoustic Emission, 164, ASTM, Philadelphia (1971).
17. H. Dunegan and D. Harris, Ultrasonics, *7*, 160 (1969).
18. D. James and S. Carpenter, Journal of Applied Physics, *42*, 4685 (1971).
19. P. Gillis and M. Hamstad, Materials Science and Engineering, *14*, 103 (1974).
20. T. Imanaka, K. Sano and M. Shimizu, Crystal Lattice Defects, *4*, 57 (1973).
21. J. Duke, Master's Essay, The Johns Hopkins University, Baltimore (1976).
22. F. Higgins, Ph.D. Thesis, University of Denver, Denver (1977).
23. K. Ono, R. Stern and M. Long, ASTM STP 505, Acoustic Emission, 152, ASTM, Philadelphia (1971).
24. L. Graham and G. Alers, Materials Evaluation, *32*, 31 (1974).

25. L. Graham, S. C. 595.6 QITR, Science Center, Rockwell International, Thousand Oaks, California (1975).
26. R. Bianchetti, M. Hamstad and A. Mukherjee, Proceedings of the Second International Conference on Materials, 1625, Boston (1976).
27. M. Bassim, D. Hay and J. Canteigne, Materials Evaluation, *34*, 109 (1976).
28. R. Kline, Master's Essay, The Johns Hopkins University (1975).
29. H. Hatano, Journal of the Acoustical Society of America, *57*, 639 (1975).
30. H. Hatano, Journal of Applied Physics, *47*, 3873 (1976).
31. K Ono and J. Ucisik, Materials Evaluation, *34*, 32 (1976).
32. J. Frederick, Report 01971-2-T, University of Michigan, Ann Arbor, Michigan (1969).
33. J. Speake, Absolute Measurement of Acoustic Parameters, Characterization of Ultrasonic Equipment (ISPRA Course) (1977).
34. N. Hsu, J. Simmons and S. Hardy, Materials Evaluation (Research Supplement), *35*, 100 (1977).
35. L. Graham and G. Alers, ASTM STP 571, Monitoring Structural Integrity by Acoustic Emission, 11, ASTM, Philadelphia (1972).
36. F. Breckenridge, C. Tschiegg and M. Greenspan, Journal of the Acoustical Society of America, *57*, 626 (1975).
37. B. Maxfeld and R. Cochran, Materials Evaluation, *31*, 17 (1973).
38. C. Palmer and R. Green, Proceedings of the 23rd Sagamore Army Materials Research Conference: Nondestructive Characterization of Materials, Raquette Lake (1976).
39. R. Kline, R. Green, Jr., and C. Palmer, Jr., Journal of the Acoustical Society of America, *64*, 1633 (1978).
40. H. Lamb, Philosophical Transactions of the Royal Society, *A203*, 1 (1904).
41. L. Caignard, Reflexion et Refraction des Ondes Seismiques Progressives, Gauthier-Villars, Paris (1939).
42. L. Knopoff, Journal of Applied Physics, *29*, 661 (1958).
43. C. Pekeris, Proceedings of the National Academy of Science, *41*, 469 (1955).
44. J. Sinclair, Journal of Physics D: Applied Physics, *12*, 1309 (1979).
45. W. Pardee, Journal of the Acoustical Society of America, *70*, 100 (1981).
46. Y. Pao, R. Gajewski and A. Geranogla, Journal of the Acoustical Society of America, *65*, 96 (1979).
47. A. Ceranoglv and Y. Pao, Journal of Applied Mechanics, *48*, 125 (1981).

48. C. Scruby, J. Collingwood and H. Wadley, Journal of Physics D; Applied Physics, *11*, 2359 (1978).
49. R. Kline, R. Green, Jr. and C. Palmer, Jr., Journal of Applied Physics, *52*, 141 (1981).

Time and Frequency Analysis of Acoustic Emission Signals

M. N. Bassim and M. Houssny-Emam

Metallurgical Sciences Laboratory
Department of Mechanical Engineering
University of Manitoba
Winnipeg, Manitoba, Canada

INTRODUCTION

Once an acoustic emission signal has been detected from a structure or specimen by a transducer, it is next amplified sufficiently for analysis. The analysis generally involves establishing relationships between signal parameters and the physical variables to which the structure is subjected, such as load, stress, strain, time, etc. The amplification is usually carried out in two steps to gains up to 120 db. The signals are also passed through appropriate filters for elimination of a large part of the background noise. Depending on the application, signal analysis may be either in the time or frequency domain. For applications requiring frequency analysis, usually in the range from 0.1 to 2 MHZ, a broad band transducer is used and the filters will have a band pass of the same frequency of 0.1 to 2 MHZ. Such transducers however, have a lower signal to noise ratio than the resonant transducers used in time domain analysis. Frequency analysis is usually performed to identify the sources of acoustic emission or when the signals are of the continuous type such as those occuring during plastic deformation or in leak detection. In this case, characteristic frequency peaks are identified which may correspond to the presence of a leaking flow and may be due to the flow of a fluid, liquid or gas, through the leak.

Because of the ability of acoustic emission to provide a real-time monitoring of structures and components, the time and/or frequency analysis equipment has to perform in real-time. The choice of such equipment and technique depends on the application and costs involved. In this article, a review of the different approaches for such analysis is presented with emphasis on their suitability for a given purpose. Advanced techniques, which are in the development stage are also included since, from the previous history of acoustic emission, these developments may become in common use in the not too distant future.

Figure 1 shows the different possibilities for time and frequency analysis of acoustic emission signals. In the time domain, the measured parameters include cumulative number of events, total count, event rate, count rate, energy of the acoustic emission signals, maximum amplitude, root mean square of the amplitude, amplitude, distribution and pulse width. In applications involving the use of more than one acoustic emission channel for flaw location, the different arrival time from the source to the transducers is used for determination of the position of the emitting source. New time domain analysis techniques involve pattern recognition

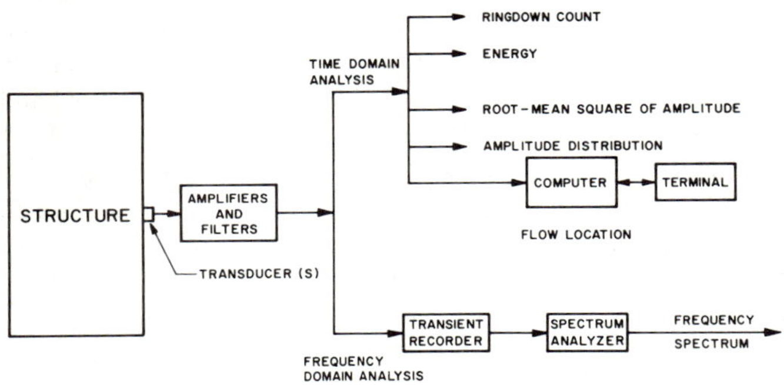

Figure 1. Block diagram showing the different parameters measured in the time and frequency domains.

and auto and cross correlation. The frequency domain analysis involves use of Fast Fourier Transform Analyzer (FFT), spectrum analyzers as well as possibly the use of digital computers. Thus appropriate interfaces, including transient recorders, or digital oscilloscopes are used in this analysis.

1. TIME DOMAIN ANALYSIS

1.1 Ringdown Count and Count Rate (Number of Threshold Crossings)

Perhaps the most commonly used technique of time domain analysis is the determination of the ringdown count and count rate. These are related to the cumulative damage and the rate of occurrence of the damage respectively. After amplification, the signal is passed in an electronic counter which measures the number of times the signal exceeds a given threshold voltage. Figure 2 illustrates this procedure and shows the method of evaluation of the total count and count rate of an acoustic emission signal. These two parameters, as well as counting the

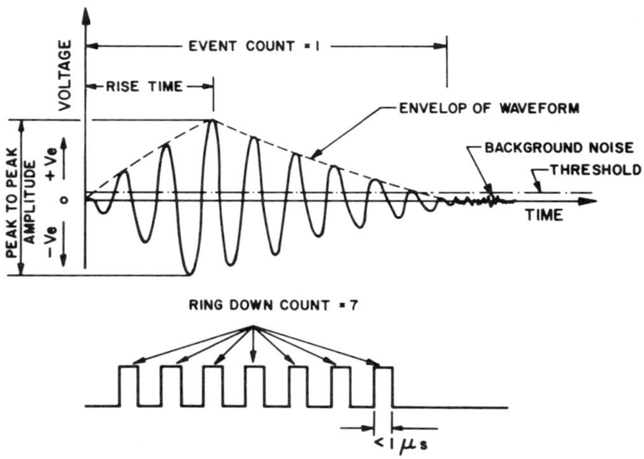

Figure 2. Ringdown counting of an acoustic emission event.

number of acoustic emission events and the rate of occurrence of these events during a test, have been correlated to the applied load, strain, number of cycles in fatigue testing and to the stress intensity factor in fracture studies.

Examples of the correlation between the total count and count rate with these physical variables are abundant in the literature. Thus, acoustic emission has been used to study the plastic deformation and yield point phenomena in metals and alloys,[1-5] fracture and crack propagation, where Dunegan *et al.*[6] and others[7,8] have proposed a relationship between the total count N_t and stress intensity factor K_I of the form

$$N_t = A\, K_I^m \tag{1}$$

where m and A are constants which depend on the material. The count rate N has been related to the stress intensity factor in fatigue[9-11] by equations of the form

$$N = B(\Delta K)^n \tag{2}$$

Here, B and n are also material constants which are related to the constants in Paris Law for fatigue crack propagation. The total count and count rate measurements were used to study corrosion,[12,13] stress corrosion cracking,[14,15] hydrogen embrittlement,[16] mechanical and fracture behavior of ceramics[17-19] as well as fiber composites.[20-23]

1.2 Energy Measurements

Despite its wide use, ringdown counting is only indirectly related to the physical characteristics of the event at the transducer level and hence to the nature of the emission source. Another time domain approach is the measurement of the energy of the acoustic emission signal. Since acoustic emission is attributed to the release of energy in the material due to changes in the state of stresses, measurement of the energy content of the signal is believed to be an indication of this energy release. A circuit for measurement of the energy was proposed by Beattie and Jaramillo[24] and a description and significance of such measurements were presented by Harris and Bell.[25] A basic circuit for determination of the energy is shown in Figure 3. It consists in measuring the area under the envelope of the square of the signal using a digital integrator.

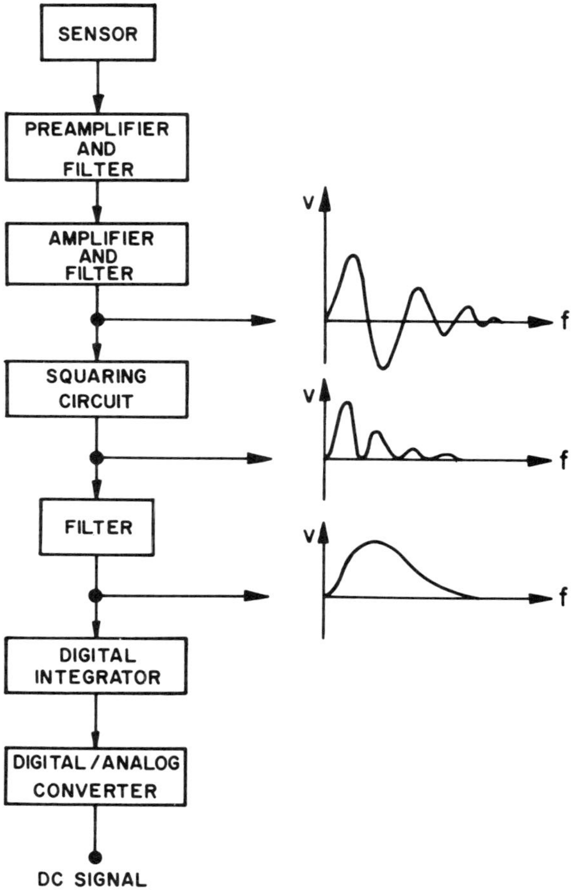

Figure 3. Block diagram for measuring the energy of acoustic emission signals (25).

Another approach in determination of the energy of the signal is by measuring the root-mean square (RMS) of the amplitude using an RMS voltmeter. For a continuous signal, such as that from a leak, with a constant amplitude V_0, the energy rate is proportional to the square of the RMS voltage. RMS measurements are very suitable for continuous signals since, in this case, no threshold voltage is imposed. The background noise, which is generally of a constant amplitude will have an RMS value which remains constant and any variation of the RMS above this value will be indicative of emissions occurring from the material tested.

The use of RMS voltage for signal analysis has been reported by Hamstad and Mukberjee[26] and by others[27] for tensile testing of aluminum alloys while Bassim *et al.*[28] have correlated the RMS voltage with the stress intensity factor during fracture of steels. They observed that the RMS value reaches a maximum as the load-load point displacement curve shows deviation from linearity during fracture testing. This behavior is shown in Figure 4 for compact tensile specimens of A36 steel.

1.3 Amplitude and Amplitude Distribution

The amplitude of the acoustic signal is indicative of the intensity of damage in the material which causes a signal to occur. It is however greatly affected by the position of the transducer with respect to the source as well as by the response of the transducer. The response to a transducer is affected by its physical characteristics and considerable effort in transducer calibration have been reported in recent years. Means of calibration presently used involves the fracturing of a glass capillary or lead pencils[29–31] as well as the continuous noise from a jet of helium on the surface of a plate.[32–33]

The measured amplitude of a signal is also a function of the position of the transducer with respect to the emitting source. Wave propagation mechanisms which are function of the propagating medium and its geometry greatly influence the form of the signal as it reaches the transducer. Bolin[34] has proposed

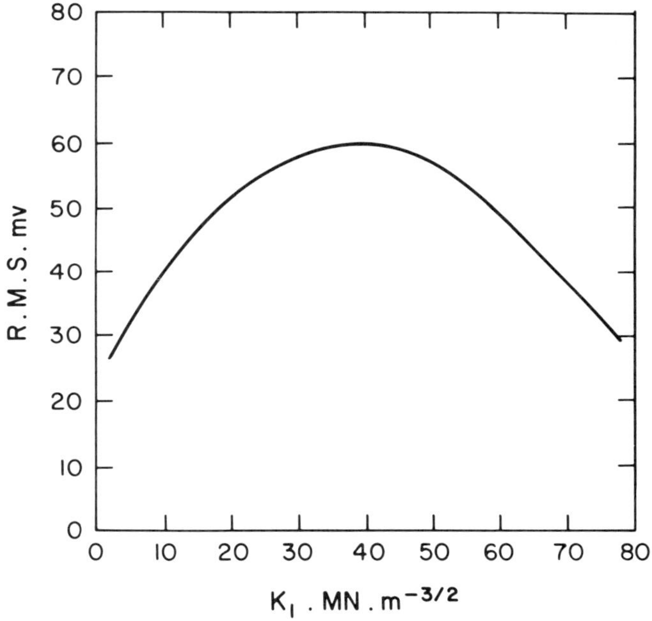

Figure 4. The variation of the root mean square of the amplitude with the stress intensity factor during a fracture test on A36 steel (37).

a model for estimation of the amplitude of the signal due to an inelastic disturbance in the material based on the analysis of Hill and Stephens[35] and Chretien[36] has studied the form of the signal at the transducer after propagation of a signal caused by an electric discharge in an aluminum bar. Bassim *et al.*[37] also studied the waveform, including the amplitude as a function of distance between a simulated source, namely the fracture of a lead pencil and have concluded that the response at the transducer is greatly affected by distance and geometry of the propagating medium as well as by the type of transducer used, whether it is a resonant or broadband transducer.

Amplitude distribution has received considerable attention as a time-domain analysis technique.[38–40] It is thought to be a characteristic feature of the material as well as the deformation mechanism. An important problem facing the user of this technique is that of attenuation in the material. Nakamura and Veach[41,42] have developed a series of spacial filtration steps to detect extremely low acoustic emission signal in a heavy background noise environment found in aircraft structural tests. Ono[43] developed a numerical amplitude distribution analysis based on Rice's theory of random signal analysis.[44] He showed that the bandwidth in acoustic emission instrumentation affected the amplitude distribution and that this analysis may be used to clarify the nature of the acoustic emission sources.

The amplitude distribution as a function of time in slow crack growth of a ceramic material, namely alumina, was investigated by Graham and Alers.[45] Their findings are shown in Figure 5a. Figure 5b gives the amplitude as a function of

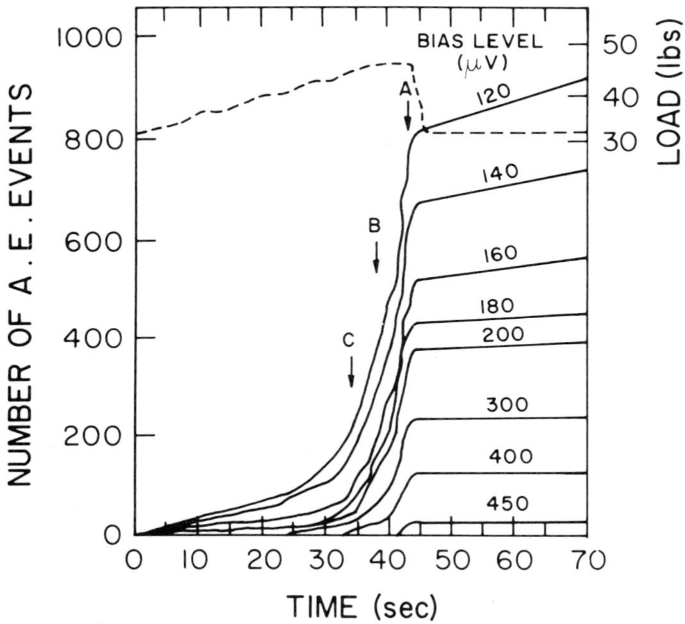

Figure 5a. Amplitude distribution of acoustic signals as a function of time for slow crack growth in alumina (45).

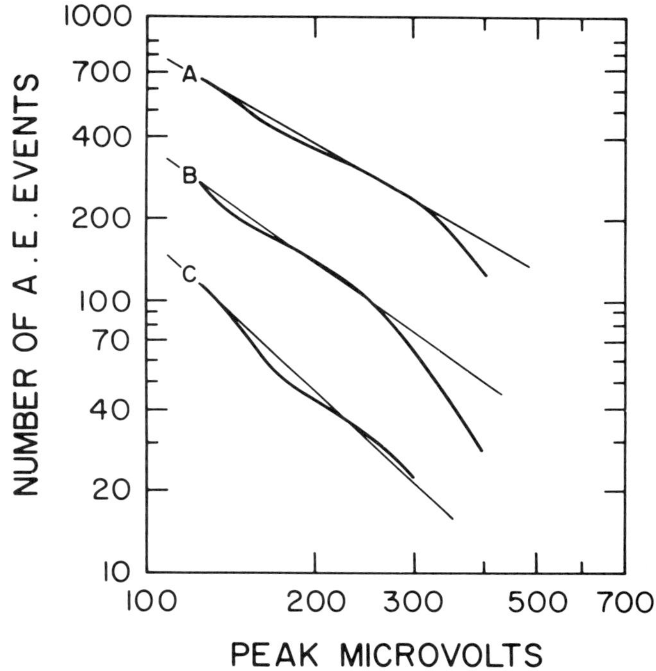

Figure 5b. Amplitude distribution of acoustic signals for three different times during slow crack growth in alumina (45).

number of events for a given time of loading. Both figures show a trend towards higher amplitudes as faster crack growth rates are obtained and corresponds to similar findings by Bassim *et al.*[28] for crack propagation in ductile steels.

A more general approach involves waveform analysis of the acoustic signal. The waveform for a signal with high frequency components is given by[46]

$$\sigma(t) = \sigma_o e^{-(t/T)^2} \tag{3}$$

where $\sigma(t)$ is the stress wave causing an emission, σ_o is the maximum amplitude of this stress wave, 2T is the pulse length and t is time. The pulse length, in this case, varies between 0.02 to 0.2μ sec. In signals due to brittle fracture, another expression has been developed:[31]

$$\sigma(t) = \frac{\sigma_o}{\sqrt{\pi}} \int_{-\infty}^{t} e(-t/T)^2 \, dT \qquad (4)$$

A value of pulse length of 0.6μ sec[47] is obtained in this case which accurately describes the observed frequency spectrum of such signals.

1.4 Flaw Location with Acoustic Emission

The most important application of acoustic emission as a nondestructive testing technique involves flaw location in real-time. For this application, more than one acoustic channel are used. The minimum number of channels are two for location of flaws along a straight line, while for special flaw location on vessels and other structures, several transducers are used which are positioned in geometric arrays. Figure 6 is an example of transducer distribution for testing a pressure vessel.

Chretien[48] has proposed two criteria for use of acoustic emission in flaw location: that the signal from the source is very rapid and that such a source has a much smaller volume than the dimensions of the structure. Based on such conditions, the continuous emissions due to plastic deformation are not suitable for flaw location, while the burst emissions from crack initiation and propagation fulfill these two criteria.

In linear flaw location, two transducers are positioned on the structure as shown in Figure 7. The signal from the source arrives at the two transducers with different arrival times. The distance x, from the source to mid point between the transducer is given by

$$x = \ell(t_a - t_b)/2 \qquad (5)$$

where t_a is the time of arrival to transducer a, t_b is the time of arrival to transducer b and ℓ is a constant describing the speed

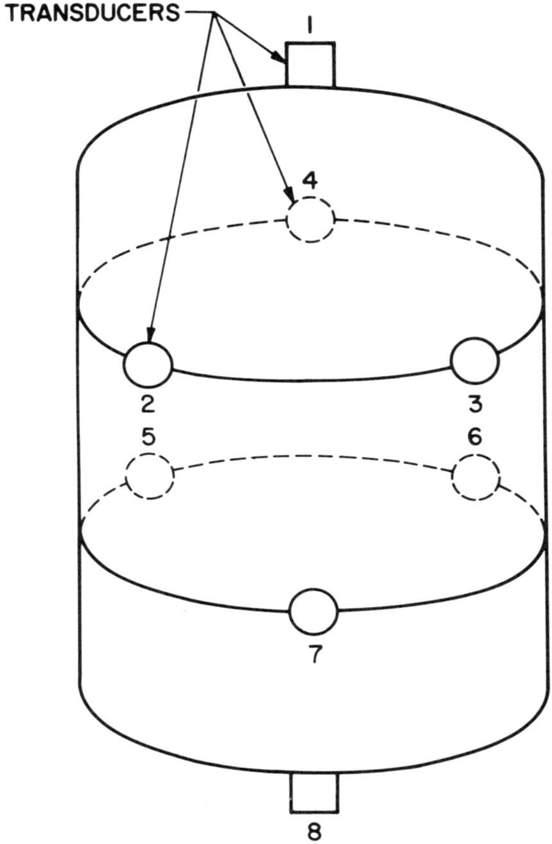

Figure 6. Position of transducers in flaw location test on a pressure vessel.

of the wave and its attenuation in the material. It is determined experimentally by a calibration procedure. The parameter $t_a - t_b$ is determined electronically using a clock.

Another recent approach for linear flaw location is by the use of the technique of cross correlation.[49,50] The delay in the

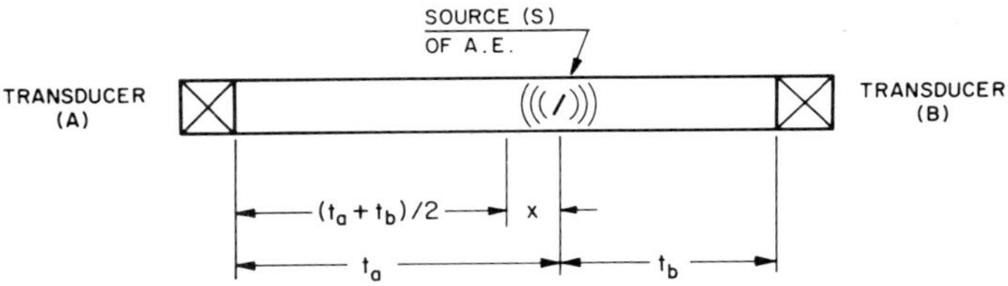

Figure 7. Linear localization using two acoustic emission transducers.

cross correlation function is thus related to the position of the source with respect to the two transducers.

In spacial triangulation using several transducers, it was shown that arrays of equilateral triangles constitute the most favorable configuration.[48] The different times of arrival to the transducers in the triangle are used to define hyperbole, the intersection of which locates the source of acoustic emission.[51,52] Figure 8 shows the intersection of two hyperbole

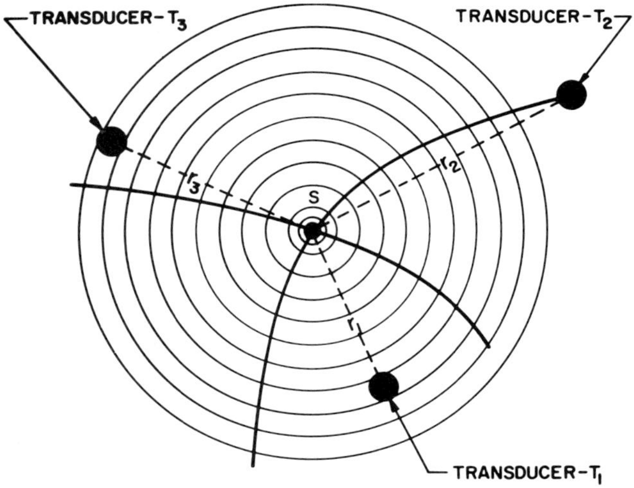

Figure 8. Spacial localization using three transducers. The figure shows the intersection of the two hyperbole defining the source of acoustic emission (53).

locating the source S. Thus, a simple set of equations of the form[53] are obtained:

$$r_3 - r_2 = k_2 = V(t_3 - t_2) \qquad (6)$$

$$r_2 - r_1 = k_1 = V(t_2 - t_1) \qquad (7)$$

and

$$r_1 - r_3 = k_3 = V(t_1 - t_3) \qquad (8)$$

In these equations, V is the velocity of sound in the material and $t_3 - t_2$, for example, is the difference in time of arrival to transducers 3 and 2 respectively.

For long time or continuous monitoring of structures, Hutton and Skorpik[54] have proposed the use of digital memories to keep the acoustic emission information until such time when analysis is necessary.

Another technique used in conjunction with multi-channel flaw location is the master-slave procedure. It involves software programming of the data in such a way that some transducers are designated as master while others are slave. This technique is commonly used to reject high background noise from machinery situated outside the area of interest. Signals triggering the slave transducer before the master transducer will be rejected, while sources of emission triggering any of the master transducer first are accepted and analyzed. A schematic diagram of this process is shown in Figure 9. Horak and

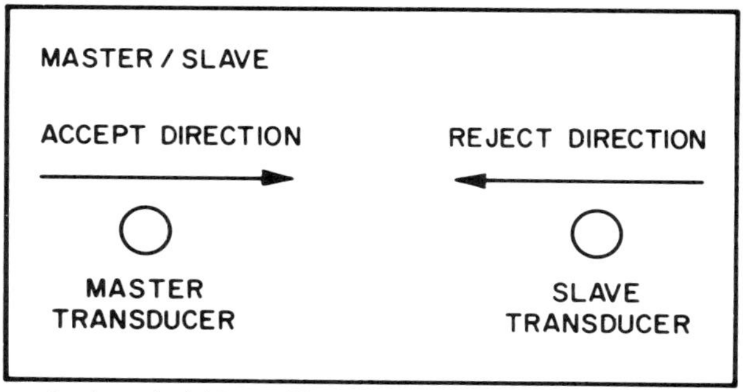

Figure 9. A typical Master/Slave transducer combination for discrimination of background noise.

Weyherter[55] have used such a system for monitoring fatigue cracking of airplane structures in a noisy environment, while Veach and Nakamura[42] used the rise time of the signal as the parameter for descrimination of the background noise by assuming that the background noise will have a longer rise time than the real acoustic emission signals. Tobias[56] describes a software program for performing flow location using a master-slave transducer set-up.

Flow location has been reported during hydrostatic testing of pressure vessels,[57,58] segments of pipeline[59] as well as in other applications including post-weld monitoring.[60] To date it appears to be the most promising application of acoustic emission as a nondestructive testing technique.

2. FREQUENCY DOMAIN ANALYSIS

While time domain analysis is more commonly used, particularly in real time applications of acoustic emission, frequency domain analysis can also be applied to identify the sources of emissions in materials and structures. Presently, however, the use of such an analysis is limited to comparative studies of the different spectra and to establishing qualitative relationships between spectra of acoustic signals and the physical processes which cause emissions to occur, such as plastic deformation, crack propagation, leak emissions . . . . etc.

The frequency content of the signal is greatly affected by the propagation medium which causes attenuation, reflections and mode conversion as well as by the transducer and the electronic system. The spectrum of the signal after simplification may thus contain characteristic frequencies of the sources as well as other frequencies which exist due to these above mentioned factors. Only in very simple cases of propagation medium geometries and with special transducers have quantitative predictions of the acoustic emission frequency content been made when using artificial sources with known input characteristics.[61–65]

2.1 Basic Principles

The basic principles of frequency analysis have been treated in several textbooks of random signal analysis.[66–68] Namely, the signal at the source can be represented in the time domain as x(t), while the observed signal after undergoing detection and amplification is y(t). The effect of the system is represented by $h(\tau)$ and is defined as the system impulse response when $x(t) = \delta(t)$. This is shown in Figure 10. The frequency response function H(f) where f is the frequency is given by the Fourier transform:

$$H(f) = \int_0^\infty h(\tau)\, e^{-i2\pi f\tau}\, d\tau \qquad (10)$$

$$= h(f)\, e^{-i\theta(f)} \qquad (11)$$

with the gain factor

$$H(f) = [H_R^2\,(f) + H_I^2\,(f)] \qquad (12)$$

and the phase factor

$$\theta(f) = \tan^{-1}\,[H_I(f)/H_R(f)] \qquad (13)$$

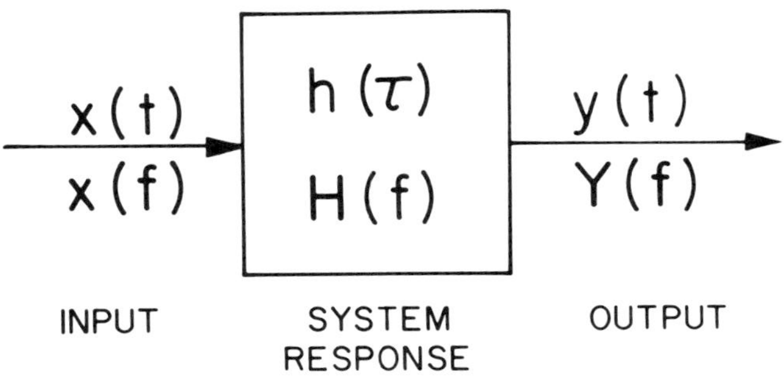

Figure 10. Schematic representation, in the time and frequency domains, of the system response to an input signal.

where $H_R(f)$ and $H_I(f)$ are the real and imaginary parts of the function H(f) and $i = \sqrt{-1}$.

In the time domain, y(t) is calculated from the convolution integral

$$y(t) = \int_0^\infty h(\tau)\, x(t-\tau)\, d\tau \tag{14}$$

If the response of the system is linear, then, in the freqency domain

$$y(f) = H(f)\, X(f) \tag{15}$$

where Y(f) and X(f) are the Fourier Transforms of the output and input signals respectively.

By performing a spectral analysis on the detected and amplified signal, we obtain the function Y(f). In acoustic emission studies, only the power spectrum S_y is obtained, or

$$S_y(f) = \int_{-\infty}^{\infty} R_{yy}(\tau) \cos 2\pi f\tau d\tau \tag{16}$$

with $R_{yy}(t)$ the auto-correlation function.

$$R_{yy}(\tau) = \lim_{T\to\infty} \frac{1}{T} \int x(t)\, x(t+\tau)\, dt \tag{17}$$

and the phase of the signal is not usually used.

From equation 15, only Y(f), which is the Fourier transform of the output signal, is determined readily. The impulse at the source X(f) is not known but is believed to be an impulse function with a very rapid rise time. H(f) can be determined for cases involving a simulated impulse function input using a deconvolution procedure which was described by Hsu and Hardy,[61] and by Houghton *et al.*.[69,70]

The choice of transducer and electronic equipment including amplifiers and low and high pass filters affects the system response function considerably. Figure 11 shows the response of a narrow band and a wide band transducer and filters to a square wave at the source,[71] Houghton[69] has derived the

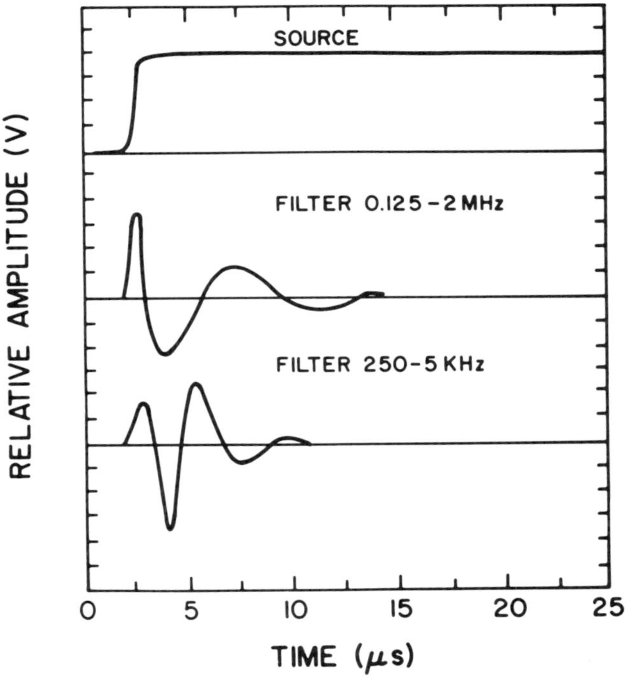

Figure 11. The response in the time domain, of two acoustic emission systems with different filters to a step function at the transducer (71).

transfer functions for these components and Blanchette[71] has obtained the response of the amplification system as shown in Figure 12.

The studies on deconvolution of the detected and amplified signal have revealed that commonly used parameters, such as ringdown counting, frequency spectra and rise-time determination can give misleading information on the input signal because of the effects of transducer and electronic equipment. This, however, has only been determined by using simulated input sources such as an electric spark and the dropping of a steel ball.[69] Also, specimen geometry, such as that of a

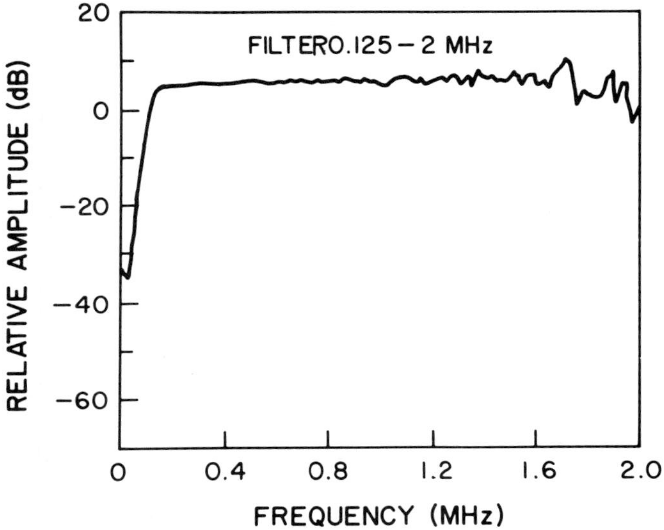

Figure 12. Frequency response of a typical acoustic emission system using a wide band transducer and filters (71).

compact-tension specimen can result in the development of constructive and destructive interferences of the acoustic signals which may affect the time and frequency analysis of these signals.[72]

Furthermore, the response of commercial piezo-electric acoustic emission transducers to the different types of waves which propagate from the source to the transducer is not clear. It was shown[61] that, while a capacitive wide band transducer is responsive to a normal displacement, that of a piezo-electric transducer might be sensitive to the time derivative of the epicenter displacement due to step function excitation. Further investigation of the behavior of piezo-electric crystals is required before conclusions can be drawn concerning the nature of the emitting source and the waveform of the signal as it reaches the transducer.

2.2 Applications of Frequency Analysis

The use of frequency spectrum analysis to identify qualitatively physical mechanisms which take place in materials has been reported. Simmons and Clough[73] have proposed a theory of acoustic emission which represents sources as stress drop rate tensors and have developed an approach for prediction of the frequency spectrum of such sources. The same authors[74] have produced acoustic emission signatures by identation using a Rockwell Hardness tester in A533B pressure vessel steels. Michaels *et al.*[75] have applied the deconvolution technique to determine the waveform of the signal at the source from records of emissions in flat plates of A533B steel. The simulated sources used were the capillary fracture source, the pencil lead fracture source and the ball impact source.

The influence of the physical processes incurred in a material on the frequency spectrum was first studied by Stephens and Pollock.[46] Other studies correlating frequency spectra with different tests include plastic deformation[76–78] and fracture of steels and other materials.[79] Recently Egle *et al.*[80] have distinguished between cracking mechanisms and dislocation mechanisms at the yield stress of nodular cast iron using frequency analysis. Figure 13 shows the different spectra obtained during compression testing of this material.

In industrial applications, frequency analysis was used by Hay *et al.*[81] to distinguish between the different stages of starting up of a methane reformer furnace. The signals were recorded and analyzed. Kupciss[82] has used frequency analysis to identify leaking pressure vessel tubes in a nuclear reactor. Woodward and Harris[83] have shown that the frequency spectra from three different mechanisms in zirconium are different and that hence acoustic emission can be used to differentiate between these mechanisms. Other applications involve ceramics[84] and determination of the quality of welds.[60]

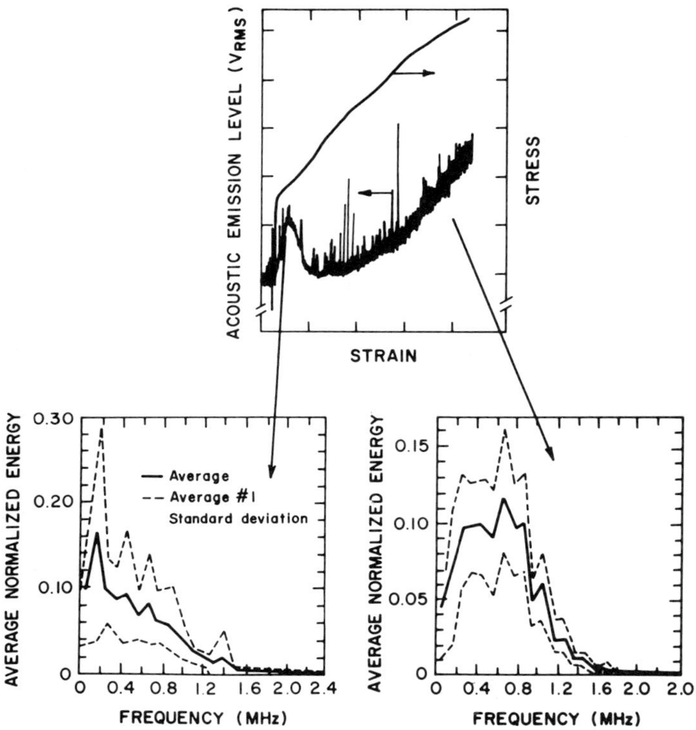

Figure 13. Frequency spectra from two sources during a compression test of nodular cast iron (80).

3. CONCLUSIONS

The different approaches in the time and frequency domain for analysis of acoustic emission signals were reviewed. In the time domain, ringdown counting remains as the most common technique and can be easily correlated to the various physical processes that the material or structure undergoes under an applied external solicitation. Accurate flaw location is the most important objective in application of acoustic emission to industrial structure to determine the presence and severity of existing flaws. Frequency analysis of the signals helps providing qualitatively an insight to the emitting sources. However, the effect of the material, transducer and electronic equipment on the frequency spectrum still remains a major problem in providing a qualitative analysis of the frequency spectrum and hence acts as a limitation to the industrial application of frequency analysis.

REFERENCES

1. K. N. Tandon and K. Tangri, Mat. Sci. Eng., *20*, 47 (1975).
2. D. R. James and S. H. Carpenter, J. Appl. Phys., *42*, 4685 (1971).
3. P. P. Gillis and M. Hamstad, Mat. Sci. Eng., *14*, 103 (1974).
4. T. Ingham, A. L. Stott and A. Cowan, Int. J. Pressure Vessels and Piping, *2*, 31, 1974.
5. M. N. Bassim and M. Veillette, Mat. Sci. Eng., *50*, 285, 1981.
6. H. L. Dunegan, D. O. Harris and C. A. Tatro, Eng. Fract. Mech., *1*, 105, 1968.
7. I. G. Palmer and P. T. Heald, Mat. Sci. Eng., *11*, 181 (1973).
8. J. Masounave, J. Lanteigne, M. N. Bassim and D. R. Hay, Eng. Fract. Mech., *8*, 701 (1976).
9. D. O. Harris and H. L. Dunegan, Exp. Mech., *14*, 71 (1974).
10. T. C. Lindley, I. G. Palmer and C. E. Richards, Mat. Sci. Eng., *32*, 1 (1978).
11. A. C. E. Sinclair, D. C. Connors and C. F. Formby, Mat. Sci. Eng., *28* (1977).
12. F. Mansfeld and P. J. Stocker, Corrosion, *35*, 541 (1979).
13. T. W. Rettig and M. J. Felsen, Corrosion, *32*, 121 (1976).
14. B. Cox, Corrosion, *30*, 191 (1974).
15. H. H. Chaskelis, W. H. Callen and J. M. Krafft, NRL Memorandum Report 2608 (1973).
16. A. S. Tetelman, "Effects of Chemical Environment on Fracture Processes", C. J. Osborn and R. C. Gifkins eds. Proc. 3rd Tewksbury Symposium on Fracture, 1974, pp. 78-97.
17. A. G. Evans and M. Linzer, J. Am. Ceramic Soc., *56*, 575 (1973).
18. A. G. Evans and L. J. Graham, Acta Metallurgica, *23*, 1303 (1975).
19. D. J. Lloyd and K. Tangri, J. Mat, Sci., *9*, 482 (1974).
20. R. T. Anderson and T. J. DeLacy, Metal Progress, *102*, 88 (1972).
21. D. O. Harris, A. S. Tetelman and F. A. Darwish, "Acoustic Emission", ASTM STP *505*, 238 (1972).

22. R. L. Mehan and J. V. Mullin, J. Composite Materials, *5*, 266 (1971).
23. C. W. Owston, in "Non-Destructive Testing and Inspection Applied to Composite Materials and Structures", Proc. of 32nd AGARD Structures and Materials Panel Meeting, London (1971).
24. A. G. Beattie and R. A. Jaramillo, Rev. Scientific Instruments, *45*, 352 (1974).
25. D. O. Harris and R. L. Bell, Exp. Mech., *17*, 347 (1977).
26. R. Bianchetti, M. A. Hamstad and A. K. Mukherjee, in Proc. of 2nd Int. Conf. on Mech. Behavior of Materials (ICM II), (1976) pp. 1625-1630.
27. A. E. Brown and R. G. Liptai, "Acoustic Emission", ASTM STP *505*, 318 (1972).
28. M. N. Bassim, D. R. Hay and J. Lanteigue, in Proc. of 2nd Int. Conf. on Mech. Behavior of Materials (ICM II) (1976), pp. 1636-1640.
29. N. N. Hsu, "Acoustic Emission Simulator, U.S. Patent 401804 (1976).
30. N. N. Hsu, "A Mechanical AE Simulator for System Calibration and Waveform Analysis", 16th Meeting of the U.S. Acoustic Emission Working Group, Williamsburg, Va. (1976).
31. N. N. Hsu, J. A. Simmons and S. C. Hardy, Materials Evaluation, *35*, 100 (1977).
32. S. L. McBride and T. S. Hutchison, Can. J. Phys., *54*, 1824 (1976).
33. G. A. Green and P. Dingwall, NDT International, 175 (1978).
34. L. Bolin, Ultrasonics, *17*, 67 (1979).
35. R. Hill and R. W. B. Stephens, Acoustica, *31*, 224 (1974).
36. N. Chretien, Ultrasonics, *16*, 69 (1978).
37. M. N. Bassim, J. P. Panet and D. R. Hay, Proceedings of Ninth World Conference on Nondestructive Testing, Melbourne, Australia (1979).
38. J. Holt and T. W. M; Evans, "Measurement and Interpretation of the Amplitude Distribution of Acoustic Emission Signals", Laboratory Note No RD/L/N 40/76. Central Electricity Research Laboratories, Leatherhead, Surrey, England (1976).
39. A. A. Pollock, Nondestructive Testing, *6*, 264 (1973).
40. Y. Nakamura, C. L. Veach and B. O. McCauley, "Acoustic Emission", ASTM STP *505*, 164 (1972).
41. Y. Nakamura, Materials Evaluation, *29*, 8 (1971).
42. C. L. Veach and Y. Nakamura, "Development of Acosutic Emission Monitoring Techniques, Research Summary ARR-15 General Dynamics, Fort Worth, Texas (1971).
43. K. Ono, Materials Evaluation, *34*, 177 (1976).

44. S. O. Rice, "Mathematical Analysis of Random Signals", in Selected Papers on Noise and Stochastic Processes, Ex, Ed., Dover, New York, 185 (1954).
45. L. J. Graham and G. A. Alers, "Fracture Mechanics of Ceramics", Bradt, Hasselman and Lange, eds, Plenum, 175 (1974).
46. R. W. B. Stephens and A. A. Pollock, J. Acoustical Soc. Am., *59*, 904 (1971).
47. Y. Blanchette, J. I. Dickson and M. N. Bassim, Engineering Fracture Mechanics (1982).
48. N. Chretien, Bulletin d'Informations Scientifiques et Techniques, Commissariat a l'Energie Atomique, No. 211, 85 (1976).
49. B. F. Cron and C. H. Sherman, J. Acoustical Soc. Am. *34*, 1732 (1962).
50. J. Grabec, Ultrasonics, *16*, 11 (1978).
51. N. O. Cross, L. L. Loushin and J. L. Thompson, Acoustic Emission" ASTM STP *505*, 270 (1972).
52. M. R. Compton, "A Real-Time Software Operating System for a Computer Controlled Acoustic Emission Flaw Monitor", Batelle Pacific Northwest Laboratories, BNWL-SA-4501, 1972.
53. T. Licht, "Acoustic Emission", Brüel & Kjaer Technical Report, 1980.
54. P. H. Hutton and J. R. Skorpik, Materials Evaluation *35*, 55 (1977).
55. C. R. Horak and A. F. Weyhreter, Materials Evaluation, *35*, 59 (1977).
56. A. Tobias, "A.C.E.M.A.N. – A PDP 11 Operating System for Acoustic Emission Analysis", Central Electricity Generating Board RD/B/N3188 (1974).
57. B. H. Schofield, "A Study of the Applicability of Acoustic Emission to Pressure Vessel Testing, AFML-TR-66-92, Lessels and Associates, Waltham, Mass. (1966).
58. "Reactor Pressure Vessel Integrity: In-Service Examination", Nuclear Engineering International, *19*, 757 (1974).
59. P. H. Hutton and J. B. Vetrano, "Crack Detection in Pressure Piping by Acoustic Emission", Battelle Northwest BNWL-1315-2 (1970).
60. M. N. Bassim, D. R. Hay and M. Talerman, Materials Evaluation, *37*, 49 (1979).
61. N. N. Hsu and S. C. Hardy, in "Elastic Waves and Non-Destructive Testing of Materials, Y. H. Pao ed., ASME publishers, 85 (1978).
62. Y. H. Pao, Elastic Waves and Non-Destructive Testing of Materials, AMD 29, Y. H. Pao ed., ASME, 1-7 (1978).

63. W. Sachse and A. N. Ceranoglu, Ultrasonics International 1979, Conference Proceedings, IPC Science and Technology Press Ltd., Guilford, England, 138 (1979).
64. Y. H. Pao, R. J. Gajewski and A. N. Ceranoglu, J. Acoustical Soc. Am., *65*, 96 (1979).
65. H. A. Crostack, Ultrasonics, *15*, 253 (1977).
66. J. S. Bendat and A. G. Piersol, "Random Data Analysis and Measurement Procedures", John Wiley & Sons Inc., NY (1971).
67. D. F. Mix, "Random Signal Analysis", Addison-Wesley Publishing Co. (1969).
68. "Digital Signal Processing, L. R. Rabiner and C. M. Rader, eds., IEEE Press (1972).
69. J. R. Houghton, M. A. Towsend and P. F. Packman, J. Acoustical Soc. Am., *61*, 859 (1977).
70. J. R. Houghton, M. A. Towsend and P. F. Packman, Exp. Mech. *18*, 233 (1978).
71. Y. Blanchette, M. Sc. Thesis, Ecole Polytechnique, Montreal, Quebec, Canada (1980).
72. F. Hamel, M. N. Bassim and J. P. Bailon, Ultrasonics, *17*, 125 (1979).
73. J. A. Simmons and R. B. Clough, "Theory of Acoustic Emission", in Dislocation Modelling of Physical Systems, Proceedings of the International Conference, Florida 1980, pp. 464-497.
74. R. B. Clough and J. A. Simmons, Materials Evaluation, *39*, 1026 (1981).
75. J. E. Michaels, T. E. Michaels and W. Sachse, Materials Evaluation, *39*, 1032 (1981).
76. D. M. Egle and C. A. Tatro, J. Acoustical Soc. Am., *41*, 321 (1967).
77. W. F. Hartman and R. A. Kline, Materials Evaluation, *35*, 47 (1977).
78. S. N. Carpenter and F. P. Higgins, "The Identification of Sources of Acoustic Emission during Deformation", Third Acoustic Emission Symposium, Tokyo, 288 (1976).
79. L. J. Graham and G. A. Alers, Materials Evaluation, *32*, 31 (1974).
80. D. M. Egle, C. A. Tatro and A. E. Brown, Materials Evaluation, *39*, 1037 (1981).
81. D. R. Hay, M. N. Bassim and D. Dunbar, Materials Evaluation, *34*, 20 (1976).
82. O. A. Kupcis, "Non-destructive Inspection of Pressure Tubes at the Pickering Nuclear Generating Station", Proceedings of the Third Conference on Periodic Inspection of Pressurized Components, Inst. of Mech. Eng., London (1976).

83. B. Woodward and R. W. Harris, Acoustica, *37*, 190 (1977).
84. L. J. Graham and G. A. Alers, "Frequency Spectra of Acoustic Emission Generated by Deforming Metals and Ceramics", Proc. IEEE Ultrasonics Symposium, Boston (1972).

Index